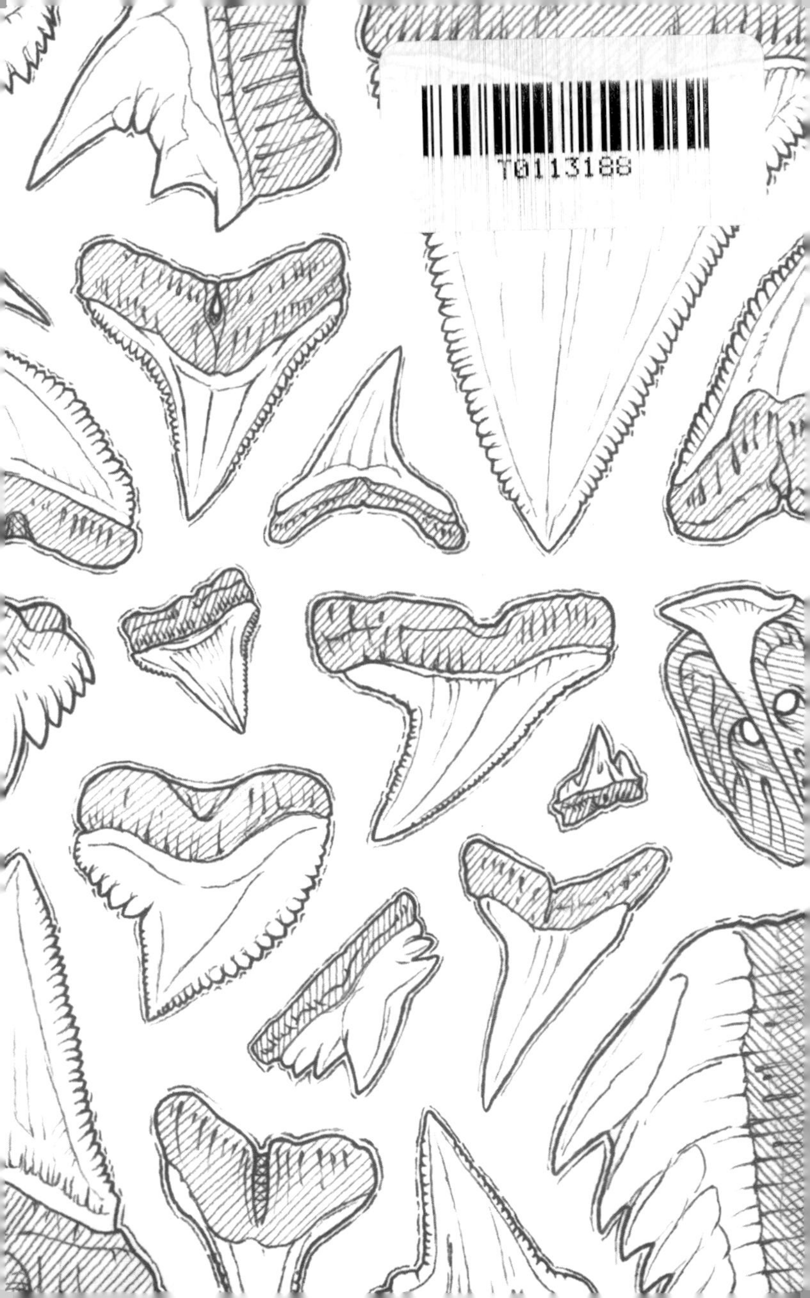

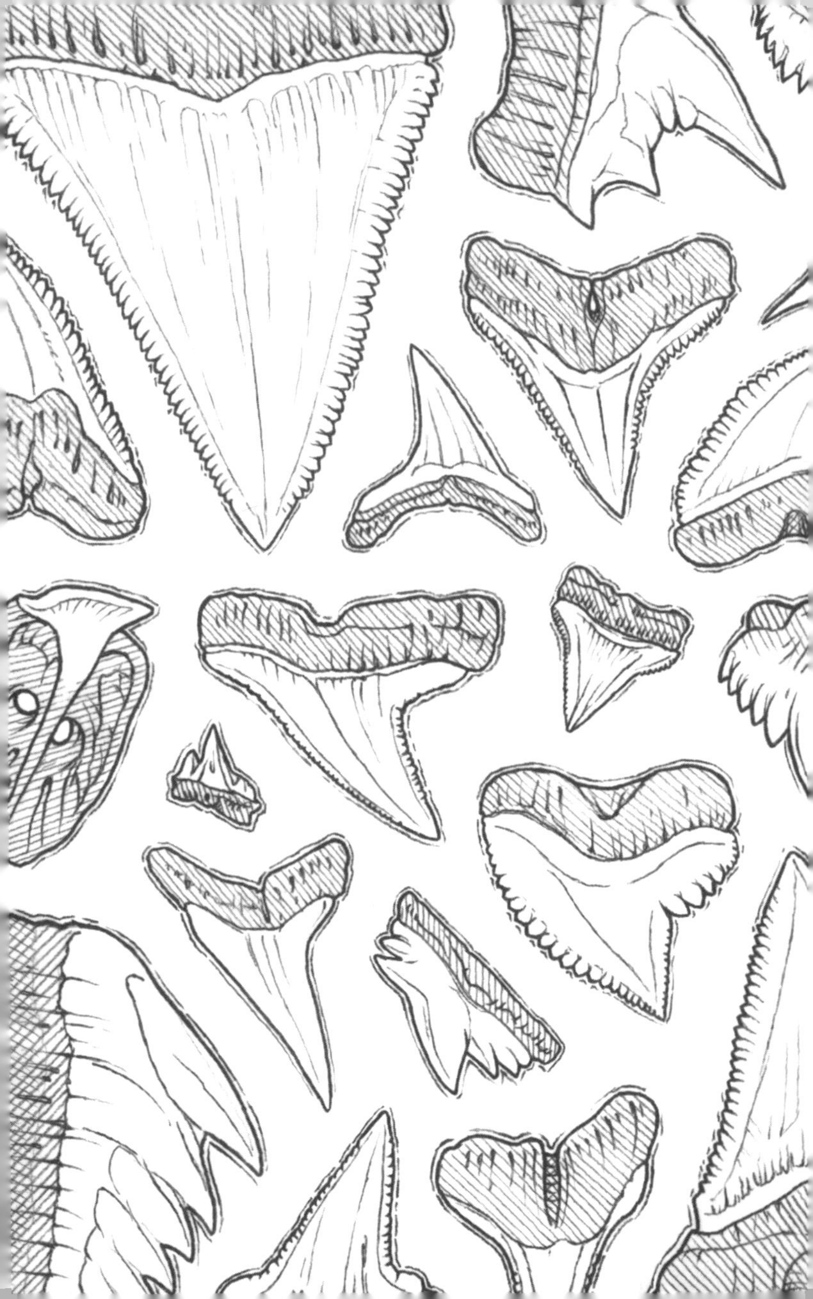

Sharkpedia

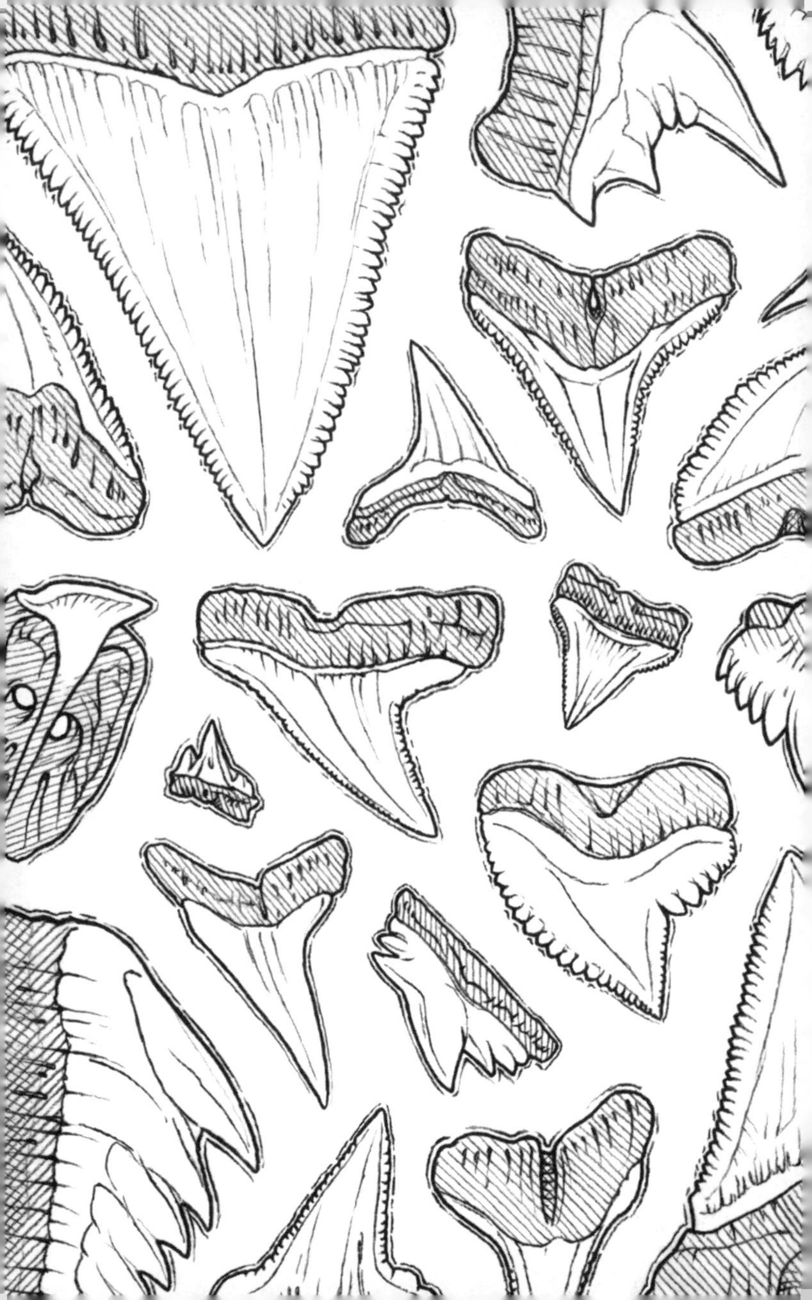

Sharkpedia

A Brief Compendium of Shark Lore

Daniel C. Abel

Illustrations by Marc Dando

PRINCETON UNIVERSITY PRESS
Princeton & Oxford

Published by Princeton University Press
41 William Street, Princeton, New Jersey 08540
99 Banbury Road, Oxford OX2 6JX

press.princeton.edu

ISBN 9780691252612
ISBN (e-book) 9780691252629

British Library Cataloging-in-Publication Data is available

Editorial: Robert Kirk and Megan Mendonça
Production Editorial: Mark Bellis
Text and Cover Design: Chris Ferrante
Production: Steve Sears
Publicity: Matthew Taylor and Caitlyn Robson
Copyeditor: Lachlan Brooks

Cover, endpaper, and text illustrations by Marc Dando

This book has been composed in Plantin, Futura, and Windsor

Printed on acid-free paper. ∞

Printed in China

10 9 8 7 6 5 4 3 2 1

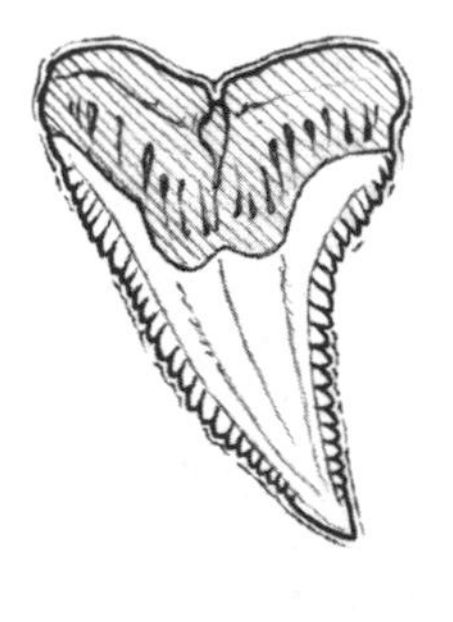

Preface

What is it about sharks that galvanizes otherwise sane and reasonable people into polar extremes at the mere mention of the word? Can you name even one other group of animals whose devotees are matched by its detractors in both quantity and passion? Snakes might come close, but until there is an annually televised Snake Week as popular as Shark Week, I'll not be convinced there are as many snake enthusiasts as shark celebrants, although sadly, many people react with similar negativity to both sharks and snakes.

Interestingly, it's very likely the same quality that divides people into shark lovers or haters—their perceived prowess as outsized, powerful predators. This perception, however, in reality applies only to a minority of shark species. While all sharks are indeed predators, even plankton-eating species like Basking and Whale Sharks, it might surprise you to learn that about two-thirds of the 541 or so (and growing) extant species of sharks are less than 3.3 ft (1 m) long. Also, more than half of all shark species live in the deep sea, defined as depths below 660 ft (200 m). Thus, streamlined, large gray beasts that live along shorelines or in the ocean's surface waters are not typical sharks, but small, brown sharks that inhabit the deep sea—the most expansive ecosystem on the planet—are.

The public thus loves or hates (or perhaps *fears*) the *least* representative sharks, such as the White Shark, Tiger Shark, or Bull Shark, while they remain oblivious to the existence of the more common species, for example, the Gulper Shark, Horn Shark, or Cuban Dogfish. Such a distinction has its consequences, some of which are not innocuous. We care most about that which we know, and thus our ignorance of deep-sea and other lesser-known sharks, even if totally understandable, translates into less attention by the media, less regulation, and less research funding.

I saw my first live shark, a majestic creature, while fishing with my older brothers as a preteen. The emergence of a disproportionately large dorsal fin from the murky, sediment-soaked depths of Deewees Inlet, just north of Charleston, South Carolina, was the first evidence that indeed we had hooked a shark, which we mistakenly called a *sand shark* (the accepted common name is Sandbar Shark), at the time the most often encountered shark in local waters. I recall my emotions then as comingled fear, excitement, awe, curiosity, and reverence. To call the experience transformational is an understatement; it launched me into a life and career trajectory of valuing and respecting nature, wanting to learn and teach about it, and, ultimately, working to preserve it. And so began my love affair with sharks.

I suspect I share these feelings with most readers of *Sharkpedia*.

Fifty-plus years later, Sandbar Sharks retain a very special place in my memory, and I still relive the same sentiments of that childhood experience whenever I see a live shark, especially while I'm snorkeling. If I am successful in writing a book that retains the essence of these feelings, adds knowledge from the thoughtful research of generations of shark biologists (plus some of my own), promotes valuing and conserving sharks specifically and the natural world in general, and does so without either insulting your intelligence or being too professorial, then I will have achieved my goal in writing *Sharkpedia*. If this book motivates you to learn more about sharks (or nature) and take action to ensure that the biodiversity of sharks and all the natural world is maintained, then I will celebrate.

One final thought: how did I select the terms included here? Having recently published two books on sharks (*Shark Biology and Conservation* and *The Lives of Sharks*), both with noted shark biologist R. Dean Grubbs, as well as a chapter on sharks in *Tooth and Claw: Top Predators of the World*, with experts Robert Johnson and Sharon Gilman, I thought I had written everything that I knew and needed to communicate about sharks. Ha! My initial list of terms that I considered essential for *Sharkpedia* surpassed three hundred, many of them unmentioned in my previous works. I quickly came to the conclusion that identifying the perfect mixture of biology, conservation, history, lore, iconic species, overlooked species, salient terminology, contemporary references, major historical figures, and art and literary references was a fool's errand; I could not write the shark equivalent of the *Oxford English Dictionary* (the *Sharkford English Dictionary*?). What survived the cutting room floor is a representative sampling of my initial list, one that serves as more of a launchpad for readers to explore their own curiosity than the definitive list of shark terms. It begins with "Adaptation" and ends with "Why Sharks Count," a chronology of sorts that tracks my own evolution of interests as a shark biologist: *be awed by what makes sharks so cool, then act to ensure that where there is historical shark habitat, there are still sharks.*

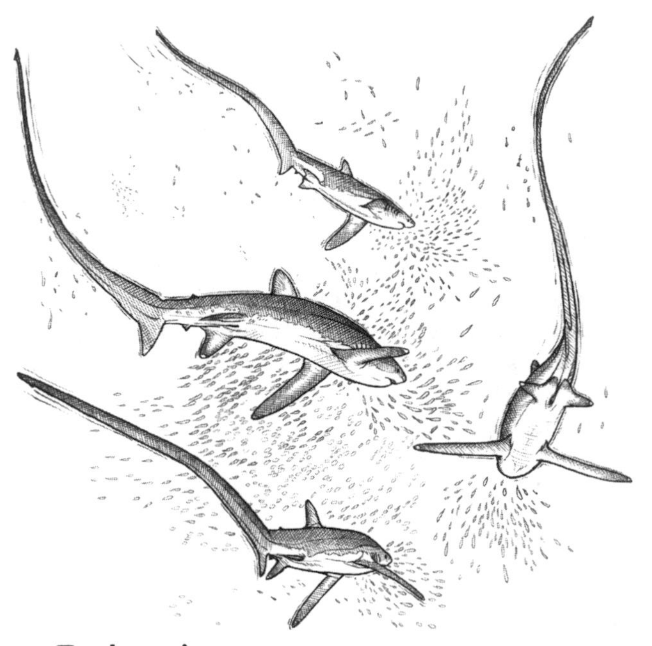

Adaptation

The nearly flawless shape of a Shortfin Mako that causes engineers and designers to drool . . . the graceful and gravity-defying aerobatics of a White Shark in mad pursuit of a Cape Fur Seal . . . the specialized, brawny jaws of the otherworldly Port Jackson Shark that enable it to crush hard-bodied prey . . . the Common Thresher's uncommonly elongated upper caudal fin lobe that herds and stuns schooling sardines . . . the ampullae of Lorenzini on the head of a Small-spotted Catshark (indeed, all sharks) that detect vanishingly small electrical currents to guide it to its prey—these are only a handful of shark *adaptations*, features that enhance the ability of these marvelous beasts to survive and that are passed on to offspring, a genetic heritage that provides the basis for their success. Adaptations come in many varieties, including anatomical, behavioral, and physiological (internal function).

It is fitting that the first entry in this book is the term that for most of us embodies our fascination with sharks, especially those adaptations that make sharks exceptional predators. Sharks and other organisms of all stripes challenge us to first identify which features in fact represent adaptations and then to understand these adaptations, that is, to determine the specific problems they solve, and the mechanisms underlying them—the ways the adaptations work. And, perhaps most meaningfully, to marvel at them. You will find all of the above adaptations, plus numerous others, in this book.

Age and Growth

Pencil scribbles on the door of a bedroom in my home record the heights of my two children's march from toddlers to teenagers—in other words, how tall they were at various ages. In essence, when ecologists refer to age and growth studies, they seek similar data, which are central to understanding a species' ecology and managing that species.

Sharks are not quite as cooperative as my children were. Fortunately, in principle, determining age and growth rate in most sharks is fairly easy and relies, at least until newer methods are adopted, on the same idea that trees deposit hard tissue periodically. In sharks, the vertebral centra, the main structural elements of the backbone, are commonly used, especially in species of coastal and open ocean zones, which have more heavily calcified vertebrae, but less so in deep-sea sharks.

What have these studies taught us? First, sharks are long-lived. A White Shark may live to more than seventy years, and a Sandbar Shark about thirty-five. The longest-lived shark known to science is the Greenland Shark, whose maximum age has been estimated to be at least 272 years (using a different method, carbon dating). One that old today would have been in its late twenties when the US Declaration of Independence was signed, and an octogenarian when Charles Darwin made landfall on the Galapagos Islands.

Sharks are also slow-growing, with comparatively late ages of maturity, and the age of maturity can also vary geographically. Maturity of Sandbar Sharks occurs at about sixteen, fifteen, and ten years for populations in the Indian Ocean, northwest Atlantic Ocean, and the Pacific Ocean off Hawaii, respectively, in all cases at lengths of less than about 5 ft (1.5 m).

So, that "trophy" 6.5 ft (2 m) Tiger Shark on the wall of the local sports bar may be only two years into its life, and six or seven years away from sexual maturity at a length of about 10 ft (3 m), all of which have been illuminated from age and growth studies.

Air Jaws

"Air Jaws" refers to the majestic aerial acrobatics of White Sharks, as well as an eponymous TV series on the Discovery Channel in the US.

It would be difficult to replicate the utter, jaw-dropping incredulity of the first time we saw a 2,200 lb (1,000 kg) White Shark hurling itself out of the water at breakneck speed, at times completely rotating midair, in pursuit of a Cape Fur Seal (or a seal decoy) in False

Bay, South Africa. How could the White Shark be so strikingly more magnificent and powerful than we had ever contemplated?

Both juvenile and adult White Sharks are known to leap, and this breaching behavior is surprisingly complex, involving different patterns of initial and follow-up strikes in which the shark may not completely leave the water. *Polaris breaches*, for example, may take the shark 10 ft (3 m) out of the water. When a shark misses its seal target, it may turn its head midair, perhaps keeping its eye on the prey for a potential do-over. Most strikes in False Bay occur within two hours of sunrise, when the murky water may limit the prey's ability to detect the countershaded shark from above but does not hinder the White Shark, which can see in low-light conditions, from sensing its potential prey from beneath.

"Air Jaws" showcases a remarkable set of behaviors made possible by the specialized, high-performance metabolic machinery of these elegant super-predators, which allows them to generate the additional muscle power needed to achieve the speed required to launch themselves. Other sharks also exhibit exciting aerial feats, which are discussed under "Jumping Sharks," but those of White Sharks are the stuff of legends.

American Elasmobranch Society

If you are a North American shark scientist, you very likely already belong to this group. If you plan on being one, you should join. The American Elasmobranch Society (AES) was founded in 1983 as a nonprofit organization that seeks *to advance the scientific study of living and fossil sharks, skates, rays, and chimaeras, and the promotion of education, conservation, and wise utilization of natural resources*. AES holds annual meetings and makes abstracts (short summaries) from the presentations at these meetings available to members. These abstracts, plus the membership directory, are invaluable to students seeking to find opportunities in shark science. AES was the first of its kind. Since its inception, similarly focused scientific societies have arisen in other regions, notably Oceania, Europe, and Brazil.

Art Depicting Sharks

Of the myriad pieces of Western art in which sharks are featured, two paintings have always stood out: Winslow Homer's *The Gulfstream* (1899) and John Singleton Copley's *Watson and the Shark* (1778). Neither, however, presents sharks favorably, which is common in most artwork on sharks until very recently. Rather than embarrass myself with an amateurish artistic interpretation, I'll leave it to you to search for these, and the fascinating stories behind them, online or—better yet—by visiting the galleries, seeing the pieces in their extraordinary majesty in person, and interpreting them yourself.

Sharks in older art, at least in Western culture, almost always have been portrayed as predators with a rapacious appetite for human flesh. Shipwreck disasters in which voracious sharks menace survivors, some of which are at least partially accurate accounts, are well represented. Consider the cover art in the French newspaper *Le Petit Parisien* depicting survivors from the schooner *Tahitienne*, a victim of a Pacific Ocean storm, rebuffing a swarm (hundreds, according to one published account) of sharks that ultimately ate nine of the eleven seamen. This type of art, even if it sensationalizes the situation and demonizes the sharks, often reflects the reality that oceanic sharks are mostly generalist, opportunistic predators for whom unfamiliar objects represent potential meals. Thus, the list of unusual items found in the digestive tracts of oceanic sharks: suits of armor, license plates, boots, wine bottles, fur coats, cans of green beans, and so on.

Another common theme in Western art is sharks attacking fishers, swimmers, and undersea divers. Artwork accompanying nineteenth- and early twentieth-century newspaper articles is particularly intriguing, even entertaining (despite the tragic stories being told).

Depictions of sharks among indigenous peoples often reflect how they considered sharks variously as gods, predators, omens, symbols, and natural resources. More so than being portrayed as objects of fear, sharks have been represented reverentially, and in a more complicated, less reactionary way in native cultures, for example as spirit animals and in rituals. For more information, research Aumakua, family gods from Native Hawaiian culture, and the sea god Dakuwaqa, from Fijian mythology, both of which are prominent in drawings and carvings.

One final note: In 1991, artist Damien Hirst created a stir with his installation entitled *The Physical Impossibility of Death in the Mind of Someone Living*—a 14 ft (4.3 m) dead Tiger Shark caught in Australia and bathed in a preservative in a glass display case. The display created a feeding frenzy among critics and the public, richly deserved in my opinion, about the questionable artistic value of the dead shark and the ethics of killing so splendid a beast for "art."

Attacks and Bites

Let's first distinguish between attacks and bites. You'd be mistaken in thinking this was a distinction without a difference: given a choice, you'd opt to be *bitten*, since bites typically cause minor injuries and are generally attributed to mistaken identity interactions.

Attacks, a word that many shark biologists think should be replaced with *serious* or *fatal bites*, typically are more injurious and may involve repeated bites. However, if a White Shark confuses you with a sea lion and mistakenly bites you, immediately recognizing that you do not taste like a sea lion and thus very likely abandoning you, you might quibble with this life-changing event being referred to as the innocuous-sounding *bite*.

In the case of mistaken-identity interactions, how do the offending sharks recognize that humans are not on their menu? Is there any overall evolutionary value to this on-the-spot assessment?

Sharks have an array of senses to locate prey, and most of the time they work exquisitely; they should, after a research-and-development period of about 450 million years. But their senses sometimes are fooled. In murky coastal waters, Blacktip Sharks, which are streamlined, swift-swimming eaters of small schooling fish, sometimes apparently confuse human hands and feet with these fish—and bite. Almost instantaneously, taste receptors in the mouth report the mistake to the brain, which replies to the jaws with a signal roughly translated into: *Blech! Spit it out! Now!* It is unknown how often mistaken identity bites occur in the lives of sharks, but the evolutionary value seems clear, to prevent swallowing prey that may be poisonous: of low nutritive value, too hard, or otherwise potentially harmful to the shark.

Both bites and attacks are low-likelihood events afforded way too much attention by the media. Given enough time, however, low-likelihood events happen. The city of Las Vegas, Nevada, would go bankrupt if gamblers rejected that point of view. Even after you enter the shark's habitat—like swimmers do at almost every beach, and in some rivers—in calf-deep water, you are safer than when you make toast, cross a street, take a shower, eat a modern diet, breathe, and so on. In the United States, air pollution causes as many as 200,000 premature deaths annually. Inhaling the dust emitted by tires and running shoes as they wear is a bigger risk to our health than sharks are. I personally know more people who have experienced direct lightning strikes than who have been bitten by a shark (and I am around people who work with sharks daily).

Globally, around one hundred or fewer unprovoked shark bites (or attacks) are reported annually, according to the Florida Museum of Natural History's International Shark Attack File. It is far from clear that this number is increasing, as media hype would have you believe. If it is, explanations, none of which have been rigorously tested, include better reporting of interactions; increases in seawater temperatures associated with climate change, which may alter shark migration patterns; more people entering shark habitat; and impacts on shark prey from overfishing, habitat alteration, and so on.

On a somewhat serious note, as a field biologist who works at sea, often in bad weather, I am aware that the odds of being struck by lightning are increased at sea. I thus try to avoid thunderstorms while at work on the water, in part because I fear the sardonic headline that might result: *Shark Biologist Validates Adage: Struck by Lightning!*

On a slightly humorous note, several years ago, an uncorroborated meme circulated on the internet that you were more likely to be killed in the tropics by strikes from random coconut falls than by sharks. Being killed by an errant coconut may be one of the few risks less likely than shark bites, although no International Falling Coconut Database collects data on coconut strikes.

Common sense and good fortune are the best deterrents to getting bitten by a shark. When my students and I see sharks, our instinct is to swim toward them. However, do not try this yourself. Instead, calmly return to your boat or the shore and leave the water. And do not enter a school of small fish (sometimes called a "bait ball"), or other areas where sharks might be attracted, like near a fishing pier.

Finally, no shark biologist would be disappointed if the media declared a moratorium on shark-bite stories and instead focused on the inherent and ecological value of sharks to their environment, and their conservation.

Batoids

Batoids, the skates and rays, are the closest relatives to the sharks, having separated from the shark evolutionary tree about 270 million years ago. From their cartilaginous skeletons to their claspers, batoids (members of the superorder Batoidea) exhibit all the defining characteristics of sharks, and are classified with sharks in the subclass Elasmobranchii.

Some shark biologists refer to batoids as *pancake* or *flat* sharks. Batoid biologists who feel slighted by this reference call their objects of study "magnificent shark pancakes of wonder," and they lightheartedly disparage sharks as "sausage rays."

Structurally, there are two main differences between the two groups. First, batoids are depressed (not emotionally, but rather in the sense of flattened), whereas

most sharks are laterally compressed and more spindle-shaped. In sharks, the five to seven bilateral gill slits are in all cases located above, or dorsal to, the well-defined pectoral fins. In rays, the five gill slits (six in one species) are located on the underside of the body, below where the broadly expanded pectoral fins, which are more commonly called wings, are attached to the body.

Over 650 of the approximately 1,250 extant species of elasmobranchs (sharks, skates, and rays) currently known to science are batoids. Iconic species include the sawfishes, mantas, stingrays, electric rays, and skates.

Most batoids are benthic, coastal inhabitants, but there are pelagic and deep-sea forms as well. Most are durophagous (eat hard-bodied prey), but some are piscivorous (fish-eaters), and others planktivorous (plankton-eaters). Members of two families inhabit low-salinity, brackish waters or even fresh water, including the humongous Giant Ray of Cambodia's Mekong River.

Many batoids are more threatened than sharks, according to the International Union for Conservation of Nature (better known as the IUCN), yet are afforded far fewer protections worldwide. Like sharks, rays have life history patterns that make them extremely vulnerable to overfishing and habitat destruction and alteration (especially for freshwater forms).

Bear Gulch

Bear Gulch sounds like a crippling disease of Grizzlies, as in, *Poor animal—looks like a case of Bear Gulch*. In actuality, it refers to the Bear Gulch limestone deposits in Montana and North Dakota, where layers of anoxic (devoid of oxygen) fine sediments, in which bacterial decomposition is limited, have exquisitely preserved remnants of an array of organisms from the Carboniferous period, about 323 million years ago. Bear Gulch was then a bay that existed for only about a thousand years. Within the approximately 90 ft (27 m) sediments, over sixty-five species of sharks, representing most of the biodiversity of the system, have been discovered. Among these was the eponymous Unicorn Shark, genus *Falcatus*, which possessed a large appendage on the head of males. One amazing find was a pair of Unicorn Shark fossils with a female grasping on to the male's head clasper. No living shark exhibits a head clasper, but the ghost sharks, or chimaeras, a small group of about fifty cartilaginous, mostly deepwater species related to the sharks and rays, do.

Bear Gulch and the Cleveland Shale of Ohio represent rare discoveries that have played disproportionately large roles in helping paleontologists piece together the phylogeny (evolutionary history) of sharks and their relatives. Prior to unearthing these two fossil-rich locations, knowledge of the evolutionary history of sharks was quite sparse, since most cartilage, the main structure material of the shark skeleton, is not heavily mineralized and does not preserve well. In Bear Gulch and the Cleveland Shale, imprints of soft tissue were well-preserved, providing unparalleled insight into the body types and ecological roles of sharks there.

Behavior

Sharks have long been considered laser-focused eating machines incapable of learning and with only primitive behaviors. While shark behaviors may not all be as sophisticated as those of many of their higher vertebrate cousins, some are surprisingly complex, and sharks can demonstrate impressive cognitive abilities.

The topic of behavior is wider than you might have envisioned. Among animals, there are numerous categories of behavior, including sexual, agonistic, social, as well as foraging (feeding), cognition (learning), predator avoidance, navigation, orientation, communication, and so on—and some of these overlap. While there are gaps, the literature on shark behavior is impressive, and is growing.

Let's consider only a few of the vast array of behaviors in the shark repertory. When the Swell Shark, a relatively small, weak-bodied benthic shark, is threatened, it self-inflates by rapidly inhaling seawater into its stomach until it is almost twice its normal size. An erstwhile predator may now reconsider attacking the bloated Swell Shark because the formerly appetizing potential prey has now become super-sized, and it may appear to be more threat than nutrition. Swell indeed.

Behavioral adaptations associated with predator avoidance, especially in smaller species or juveniles, are also common. Numerous species, such as Sandbar, Blacktip, and Bull Sharks, utilize nurseries—places that provide refuge, often from bigger sharks, and food for neonates (newborns) and juvenile sharks. Some benthic sharks, for example Horn Sharks, particularly juveniles, will hide in crevices. Others, like Pacific Angel Sharks, will bury themselves under the substrate, an action that also enables them to surprise their prey. Finally, many sharks will camouflage themselves, either through matching their backgrounds or countershading, and both may be used both to avoid predation and to conceal themselves from prey. In countershading, the upper surface is dark and the underside light. Viewed from below, a countershaded shark blends in with downwelling light, and from above, it matches the darkness of the depths.

A classic example of shark behavior is the agonistic (threat) displays of some requiem sharks, notably Grey Reef Sharks and Bull Sharks. When threatened by approaching divers, these sharks will depress their pectoral fins, arch their back, open their mouth, and stiffen their body while slowing their movement. White Sharks may behave similarly when threatened, and, being White Sharks, one-up Grey Reef and Bull Sharks by adding another dimension: convulsive, whole-body shuddering. Sharks exhibiting this behavioral suite are trying to communicate a simple message: *leave or experience the wrath of my sharp teeth*. Cornering a shark in full threat display is a potentially dangerous activity.

Studies have also shown that sharks are capable of learning. A 2022 review of the subject asserted, perhaps surprisingly, that cognition in sharks and rays was comparable with that of most other vertebrates, even birds and mammals. As with many studies of sharks, only those species that adapt well to captivity, especially in the small confines of laboratories—for example horn sharks and some carpet sharks and catsharks—have been studied, which has limited a broader understanding of learning across the group.

Mating in sharks is often violent, with males biting females or otherwise harassing them prior to copulation. In cases where there are large numbers of males competing for individual females, females have evolved ways to rebuff the libidinous hordes, or at least to discourage the advances of most of their "suitors." In their classic study, Wes Pratt and Jeff Carrier found that female Nurse Sharks escaped to the shallows, where in large congregations they were able to reject the advances of more than 90 percent of males. Peter Klimley, who studied large, mostly female, schools of Scalloped Hammerheads around the El Bajo Seamount near Mexico's Espíritu Santo Island in the Gulf of California, proposed that some females exhibited a suite of aggressive, acrobatic behaviors, such as corkscrewing, that endowed them with dominance, forcing the subordinate females to the edge of their school, and rewarding the dominant females with the front row seats of the mating arena, so to speak. Male Scalloped Hammerheads, who displayed their own behavior repertory that included the *torso thrust*, subsequently attracted and mated with the dominant females.

A few final points are worth mentioning. At least some sharks (for example, juvenile Lemon Sharks) are capable of recognizing familiar individuals of the same species, and some have also been shown to have an array of personalities, for example shyness and boldness. Port Jackson Sharks, which have been disproportionately studied because of how well they adapt to captivity, recognized jazz music and associated it with a food reward. Alas, their ear for music had limitations: they were incapable of distinguishing between jazz and blues. Dilettantes!

Bimini Biological Field Station

Pound for pound (0.45 kilogram per 0.45 kilogram, as it were), the diminutive Bimini Biological Field Station, known as "Shark Lab" to friends and local Bahamians, is among the most highly productive and important centers of tropical shark research and education globally. The lab has conducted the world's most extensive research on the ecology, behavior, life history, and physiology of the Lemon Shark, which utilizes the shorelines, the mangrove-lined lagoons, and the nearby waters from cradle to grave (although not continuously). The lab has also made contributions on other sharks, including Tiger Sharks, Bull Sharks, Great Hammerheads, and Caribbean Reef Sharks, as well as rays. The facility was founded by legendary shark biologist Samuel "Sonny" Gruber in 1990 on the westernmost island of The Bahamas, and the field station has remained active since then.

Bimini represents a hotspot for shark biodiversity, owing to its wide array of habitats and ideal location as an oasis between the deep Gulf Stream and shallow Great Bahama Bank. The lab has hosted my Coastal Carolina University Biology of Sharks course there annually since 1996, and they also offer volunteer opportunities, internships, and naturalist courses.

Biodiversity Hotspots

"Biodiversity contains the accumulated wisdom of nature and the key to its future," wrote environmental scientist Donella Meadows in 1990. *Biodiversity* generally refers to the variety of species and their population sizes in a specific ecosystem or even the entire biosphere. Ecologists calculate biodiversity indices for many ecosystems and these are then used to assess their current health or to serve as a baseline to monitor their future health. Low biodiversity is not necessarily a bad thing if the ecosystem characterized as such naturally has a low variety of species, for example the Baltic Sea (which also happens to be heavily impacted by humans).

As mostly mid-level and top predators, sharks play pivotal roles in maintaining the biodiversity of ecosystems in which they reside. They are thought to do so, like all top predators, by influencing the evolution of their prey and other organisms beneath them in the food chain, and by controlling the population sizes of these organisms, although experimental verification of these roles in sharks is sparse.

Global shark biodiversity hotspots based on species richness (the number of species) include the east and west coasts of Australia (more than 170 species), Japan, southeast Africa, Taiwan, southern Brazil, and the Southeastern United States. Within these locales, biodiversity peaks over the continental shelves—typically productive, shallow areas adjacent to continents, with abundant habitat, forage stock, and favorable environmental conditions. Other shark hotspots are regions where cold, nutrient-rich, deep water upwells to the surface, as well as seamounts—undersea oases consisting of mountains that do not breach the surface. At the Port de Pêche outdoor fish market in Casablanca, Morocco, amid a huge fleet of colorful vessels that fished near the Canary Current (an upwelling zone off northwest Africa),

I found juvenile and/or adults of several different sharks for sale, including Scalloped Hammerheads, Shortfin Makos, Small-spotted Catsharks, Blue Sharks, and Gulper Sharks (and I was detained and nearly arrested, *twice*, for performing street dissections of some of these sharks for my students and curious local residents).

Low biodiversity ecosystems (notspots?) are the open ocean—particularly in the mid-latitudes (between 25° and 40°) in both hemispheres—and polar seas, areas of lower biological productivity. Don't confuse low biodiversity with a lack of biological value (my pun above notwithstanding); these areas constitute functional ecosystems that contribute to the planet's overall biodiversity, and they serve as habitat for a small but ecologically important assemblage of sharks.

One area merits further mention, the Indo-Pacific's Coral Triangle, which covers about 2.2 million square miles (5.7 million square kilometers). Approximately 30 percent of all named shark and ray species call this area home, and its habitats include coral reefs, mangroves, and seagrass beds. Here, as in general, sharks face challenges from climate change, overfishing, and other human impacts.

Biomedical and Other Uses/Biomimicry

Sharks and their close relatives have been used by humans for over five thousand years, for food, in biomedicine, as inspiration for design, for display in aquariums, as ecotourism draws, and for a variety of shark-derived products.

One biomedical area in which sharks have been misused is as treatment for cancer. It may surprise you to learn that virtually all multicellular animal groups, from insects to whales, experience cancer. Sharks belong on this list; the idea that sharks are alone among vertebrate animals in being cancer-free is unsupported by data. Regardless of whether sharks are protected from cancer, however, there is something special about their primitive but powerful immune system that could have human clinical relevance in fighting infections and diseases.

Other biomedical uses for shark tissues include skin grafts (and wound dressing), antibacterial coverings, antibiotic and antiviral agents, bone regeneration, therapy for neurodegenerative disorders and arthritis, medical creams, and as models for biomedical research.

Shark cartilage is widely used as a health supplement, although benefits are not always supported by science. You can purchase cartilage pills, often labeled as chondroitin, which is sourced from sharks (mostly Blue Sharks), but more so from cows or pigs.

Sharks also inspire designers. Copying nature's best ideas for human uses is called *biomimicry*. Biomimetic uses of sharks include Speedo's Fastskin Swimsuit, based on the drag-reducing shapes of shark scales. These same shark scales create a texture that apparently inhibits attachment and proliferation of bacteria, and surfaces based on these scales are being developed for biomedical use. Shark swimming, and particularly the roles of fins, is being studied as a model for hulls and submersibles. And let us not forget the 1961 Mako Shark Corvette, inspired by mackerel sharks.

In addition to providing meat and fins for human consumption, sharks are caught for their liver, skin (known as *shagreen*), teeth and jaws, and other body parts. Shark liver oil, from the largest organ in sharks, is used as a lubricant in the textile industry, in cosmetics and health products, as fuel oil for lamps, and to protect boat hulls. When supplies of cod liver oil, a major vitamin A source, were cut off during World War II, shark liver oil replaced it until synthetic vitamin A was manufactured in the early 1950s. Other products from sharks and batoids include gill rakers from mantas, rostra from sawfish, and intact, preserved neonate sharks or embryos sold as curios.

Bite Force

Assumptions are ideas that we take for granted, without verification. So it is that most of us assume that the bite force of sharks is among the most powerful in the animal kingdom. Evolution doesn't waste its breath, which means that the force of shark bites is no more than is needed for grasping or crushing prey or penetrating skin and removing chunks of flesh, and these forces may or may not be as high as expected. Also, in the evolution of feeding, tooth morphology—that is, size and shape—works hand in hand with bite force to achieve the desired end: feeding.

The first scientific study of bite force in sharks, conducted in the late 1960s on adult Tiger Sharks and others in pens in Bimini, The Bahamas, found that the highest force from a single tooth was 132 lb (60 kg). If you assumed that this force applied to every functional tooth in the Tiger Shark's jaws, the total bite force would exceed three tons.

Durophagous sharks—that is, those that eat hard prey and typically possess hypertrophied jaw musculature and teeth adapted for crushing and grinding—tell a more nuanced story. The bite force of the anterior teeth of a small (3.5 lb or 1.6 kg) Horn Shark was only 26.3 lb (11.9 kg), but maximum bite force on the posterior molariform teeth was 76 lb (34.5 kg). Pound for pound, however, these are among the highest bite forces found in sharks.

The highest bite force measured in a shark was from a 423 lb (102 kg) Bull Shark, at 478 lb (217 kg), higher than that of a slightly larger White Shark. Surprisingly, perhaps, the second molar of a 154 lb (70 kg) human was close to that of the White Shark, at 292 lb (132 kg). We're gonna need a bigger car!

Bloodhounds of the Ocean

Sharks have acquired numerous epithets, of which the most common is "maneater," but two also frequently used are "bloodhounds of the ocean" and "swimming noses." These refer to the putative ability of sharks to detect vanishingly small odors, especially those associated with blood, in their environment.

There is some truth to the claim, although qualification is needed. I've seen swimmers hasten to exit from the water after a drop of blood escaped from a cut, as if every shark in the surrounding ocean instantaneously detected the red fluid and engaged its afterburners to jet to the future victim's minor injury. Detecting a scent requires that a sufficient number of specific molecules from a source capable of stimulating the shark's olfactory (odor) receptors physically contact these receptors. Thus, unless you were unlucky enough to drip blood

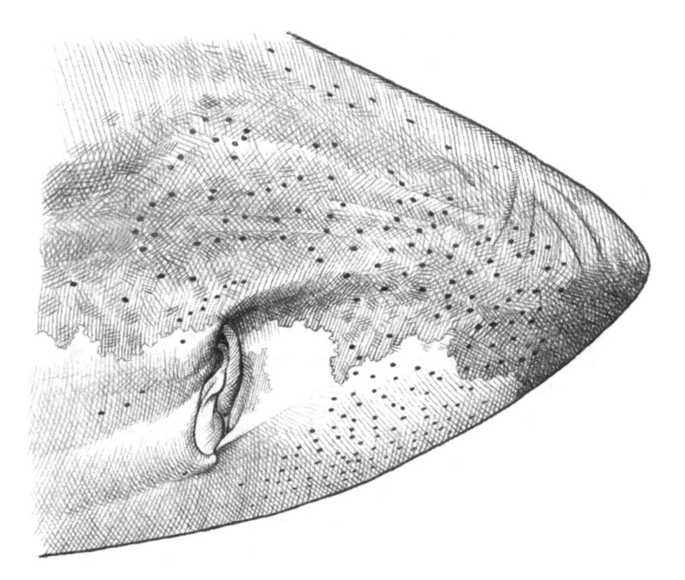

directly into the nostrils of a shark lurking uncomfortably close to you, there would be no urgency to react like you were running from Godzilla.

Now for the qualification. Sharks indeed have a prodigious sense of smell and are especially responsive to the odors of their prey. Some have even been shown to smell the difference between prey fish that have been agitated and those that are unstressed. Many sharks have a remarkable ability to track an odor corridor to its source at distances of 0.6 mi (1 km) or more. However, while this ability varies among sharks, many bony fishes have similar sensitivities, so if some sharks are bloodhounds of the ocean, so too are some bony fishes.

Blue Water, White Death

For shark enthusiasts before the early 1970s, documentaries featuring spectacular footage of sharks were sparse. *Blue Water, White Death*, a 1971 film directed by Peter Gimbel and James Lipscomb, was a game changer in that regard. In 1971, White Shark encounters were not widely publicized, cage-diving ecotourism did not exist, and scientific inquiry into the species, indeed into most sharks, was at its infancy.

To watch this documentary is to gaze through a window into the past, to simpler, more naive times when sharks were plentiful ("no way to count them," we are told about their colossal abundance during one encounter), and were even more misunderstood than now. The story, told like a travelogue, describes an expedition to the waters off Durban, South Africa, Sri Lanka, and Australia. "Sharks infest these waters," we are told about Durban, with that innocent use of the pejorative verb that conjures the image of rats swarming a landfill, as opposed to an assemblage of sharks appropriate for the area. Ron Taylor, part of the expedition, is described as the "only diver ever to see" a White Shark, which would almost be laughable now. At one point, as Oceanic Whitetip Sharks begin to actively feed on a whale carcass, someone says, "This beats the decadent night club life!" Who would dare argue with that?

The movie shines in its honesty about our ignorance of shark behavior and biology at the time. It thus is instructional to watch the film over fifty years later, when so much more is known about sharks. At the same time, the documentary has aged well, and it was groundbreaking in some of its observations about shark behavior—for example, that Oceanic Whitetip Sharks frequently bumped prey before consuming.

Other participants included documentarian Valerie Taylor, legendary underwater cinematographer Stan Waterman, and folk singer Tom Chapin.

Bobbing for Apples

Bobbing for apples is a traditional autumn activity where you attempt to grasp apples floating in a tub of water using only your teeth. Dentists might cringe at even its mention, and you might question the author's sanity for this apparent non sequitur.

If you are lucky enough to capture the Fuji or Opal apple using only your teeth, what then? Completing the bite is not trivial. And if you successfully remove a chunk, you need to repeat the entire process if you want another bite.

The task is difficult for you in large part because your upper jaw is firmly affixed along its entire length to your skull and, as such, one can't go anywhere the other doesn't also go. Plus, you must be very close to the apple, since your jaw cannot be thrust out. Here's the relevance: if the jaws of sharks were similarly configured, there would be no *Sharkpedia*, and the entry for "sharks" in the book *Dinopedia* would read: *insignificant group whose limited jaw mobility led to their demise*.

But sharks are a successful group, and this success is in part due to evolutionary advances in how their upper jaws are suspended—that is, how they are attached to supporting structures. These advances have enabled improved *cranial kinesis* (jaw mobility) and expanded gapes. Basically, the connections between the

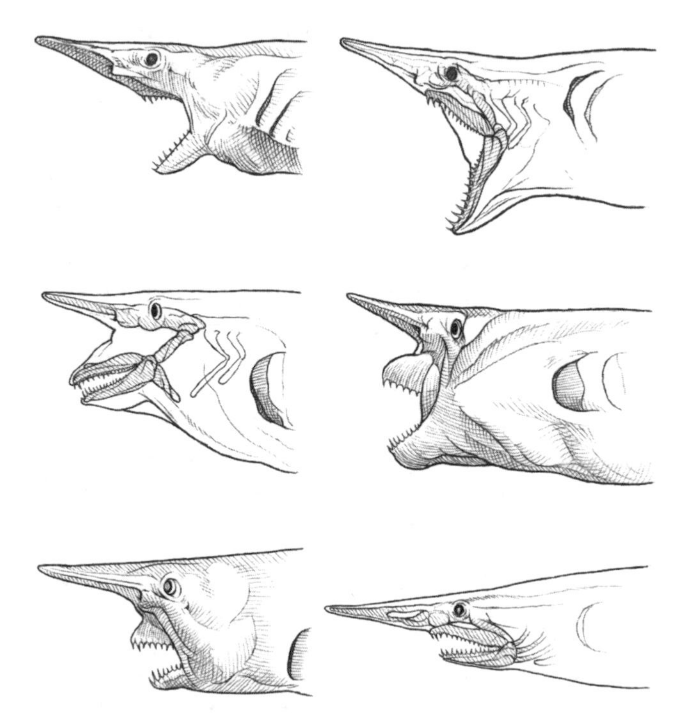

palatoquadrate (upper jaw) and the skull have loosened, and the former is connected to the latter by fibrous ligaments, with buttressing (support) provided by skeletal elements. Loosening of the jaw allows it to protrude, and thus to bite from a greater distance. It also enables an enlarged gape, since the protruding jaw, now freed from the confines of the head, can expand both vertically and laterally. Different groups of sharks have evolved varying levels of jaw mobility and types of jaw suspension, but the beasts with the most protrusible jaws are the batoids.

When you gaze in awe at videos of sharks feeding on a whale carcass, you now know both the anatomical basis underlying their success, as well as the magic of evolution.

Body Shape

A major theme in evolution is conserving energy, and moving through water takes a lot of energy, since water is about eight hundred times heavier than air and is fifty times more resistant to flowing. These same harsh characteristics, however, offer opportunities to save energy. It is therefore not surprising that sharks have evolved ways to reduce the energy expenditures of passage through a medium so hostile to movement.

For benthic sharks, simply resting on the sea floor conserves energy. For the benthic Horn Shark, continuous swimming would be especially costly in terms of energy, since its body shape (blunt head, squat body) disrupts the flow of the water in a manner only marginally different from that of a block dragged through the water, thus slowing the shark or requiring it to expend more energy. Other benthic species, such as the Nurse, Grey Bamboo, and Common Angel Sharks, are more dorso-ventrally depressed (flattened), so their shapes also discourage continuous swimming.

Pelagic sharks (those living in the water column) are more streamlined, an effective form for slicing through the water since their tapered shape minimizes the type of drag that slows the benthic sharks. Unlike the benthic species, pelagic sharks are slightly laterally (side to side)

compressed. Streamlining and lateral compression give a Crocodile Shark, for example, a big swimming advantage over a Horn Shark.

Superimposed over these broad, major differences between pelagic and benthic sharks are numerous interspecific differences. Consider head shape. The mackerel sharks, super-predators adapted to high-performance swimming, have a conical snout. Sharks that cruise the continental shelves, like the requiem sharks (Sandbar Sharks, Bull Sharks, etc.), have a blunter, wider head, less adapted for high speed but playing a more significant role in stable, horizontal swimming. The extreme laterally expanded head of hammerhead sharks favors lift and maneuverability over continuous fast swimming speed and is also a longer platform for sensory organs.

Bull Shark

One of my earliest encounters with a Bull Shark convinced me that its reputation as one of the more powerful and brutish sharks might actually be deserved, but with the addition of the adjectives *arrogant* and *graceful*, if you will excuse my being anthropomorphic. Working in a small boat in a Southeastern US estuary, we had just caught a 6.6 ft (2 m) Bull Shark on an experimental longline. Our standard approach is to secure the shark to the boat while still in the water, measure it, perhaps take a tissue sample, then insert a tag and release the shark. This beast was having none of that. As I tightened the leader attached to the hook to lift the shark's head, it gazed assertively into my eyes, and then deliberately and in slow motion moved its head from side to side once, easily severing the thick monofilament leader, like slicing through warm butter, as if to say, *Really?* It then hesitated a heartbeat to make sure I knew who was in control before swimming elegantly away. I've worked with numerous Bull Sharks since then and, while all the sharks displayed their legendary prowess, none had the impact on me of that first encounter. I guess one always remembers the first time.

Bull Sharks are members of the family Carcharhinidae, or requiem sharks. Their most prominent characteristics are a stout body; a robust, blunt, rounded snout; a large dorsal fin placed forward on the body; relatively small eyes; and, of course, an attitude. Their teeth are

broad and heavily serrated for shearing. Bull Sharks live predominantly in shallow tropical and temperate waters less than 100 ft (30 m) deep. I've seen them at beaches so shallow they almost grounded themselves. They get as deep as 490 ft (150 m). Also, only Bull Sharks and three other related shark species (plus members of three families of rays) are capable of entering into freshwater ecosystems.

Bull Sharks have earned a reputation as dangerous, especially in the developing world where the daily lives of inhabitants find them in Bull Shark habitat, particularly brackish and fresh waters into which these sharks can penetrate. Bull Sharks have been known to strike boats when they are cornered, typically after going into an agonistic (threat) display, lowering the fins and hunching their back. Finally, the widely held notion that Bull Shark aggressiveness is due to testosterone levels much higher than those of other sharks is at best unclear, lacking unambiguous substantiation.

Bycatch

Bycatch generally refers to untargeted organisms captured in fisheries. Such incidental catches may be kept or, due to low value or restrictions, discarded. Bycatch is widely considered a principal driver of population decreases for cetaceans, sea turtles, seabirds, sharks, and batoids. One thing is clear: overfishing, resulting from both incidental and targeted fishing, is the number one current threat to sharks and batoids and a much greater danger than habitat degradation, pollution, climate change, and so on (at least until these competing threats overtake it).

Unfortunately, it is hard to quantify how much of shark catch is unintentional. Moreover, some incidental catch can be valued. Even bycatch that consists of few or no sharks may be of concern in that it may include the forage base (food source) for sharks, and thus may have indirect consequences for sharks or for the ecosystems in which they reside.

Sharks that are most vulnerable to being caught as bycatch, and in shark fisheries as a whole, are those with more conservative life history characteristics (lower fecundities, longer gestation periods, later maturities, etc.), such as deep-sea sharks. One such shark is the Gulper Shark, which takes more than thirty years to mature, has four or fewer offspring per pregnancy, and has a two-year gestation period.

Virtually all marine fisheries have shark bycatch, including bottom trawls (for example, for shrimp and cod), gillnets (squid and salmon), demersal (bottom) longlines (Chilean Sea Bass, Greenland Halibut, and Hake), pelagic longlines (Mahi-mahi and Swordfish), and purse seines that use fish aggregating devices (for tunas).

Although this list will vary by location and is incomplete, sharks widely impacted as bycatch include angel sharks, Bonnetheads, dogfish, and catsharks (bottom trawls); Blue Sharks, Salmon Sharks, and juvenile Sandbar and Dusky Sharks (gillnets); Blue, Silky, and Oceanic Whitetip Sharks (pelagic longlines); Cuban Dogfish, Portuguese Dogfish, lantern sharks, and gulper sharks (demersal longlines); and Oceanic Whitetip and Silky Sharks (purse seines).

Even when commercial fishers release sharks, for some species post-release mortality may be high. For example, more than 95 percent of Cuban Dogfish caught on demersal longlines are released alive, but about half soon die. In some cases, indelicately removing the hook from the shark's mouth breaks the lower jaw.

Bycatch can be addressed in several ways. Simply tending the gear more regularly can significantly reduce mortality in some cases. On pelagic longlines, increasing the depth of hooks can reduce bycatch of Silky and Oceanic Whitetip Sharks, which spend most of their time at depths shallower than 330 ft (100 m). Fisheries can be excluded, year-round or temporarily, from areas where sharks and rays congregate or where their populations are depleted. In the US shrimp trawl industry, devices that shunt sea turtles from the nets also exclude larger sharks, for example, Blacknose Sharks. Tests are also underway to develop hooks that repel sharks by jamming their electrosensory systems, but that do not affect target species. One, called SharkGuard, was recently demonstrated to reduce Blue Shark bycatch in trials in a Bluefin Tuna fishery. It is important to note that regulations and incentives for commercial fishers to accept bycatch reduction devices and techniques vary and compliance may be very low without enforcement. It is also important to note that not all fisheries have high bycatch levels in all areas, nor are all the sharks caught as bycatch vulnerable to population depletion as a result—for example, Atlantic Sharpnose Sharks caught in shrimp trawls in the Southeastern United States.

Educating consumers so they can make ethical, sustainable seafood choices that take shark bycatch into account could also play a role, especially in wealthier areas, although getting the information to these consumers has been problematic. More research also needs to be undertaken to identify and implement additional mitigation measures. The window of opportunity is fast closing: human population size is now greater than eight billion, and demand for seafood is expanding.

Captivity and Capture Stress

As a child, observing Sandbar Sharks swimming so elegantly at the New England Aquarium not only changed the way I had considered sharks—as terrifying predators—but also set me on a trajectory to my career as a marine biologist.

While recognizing the immense ethical and environmental questions surrounding capturing, transporting, and maintaining sharks for display, I also know there are benefits, or at least the potential for benefits, of responsible husbandry. These include educating, inspiring, and promoting conservation values and action. It did in my case. Modern shark husbandry also seeks to maintain and breed species of conservation concern, so-called *insurance populations*, as hedges against population decreases in nature.

Keeping sharks in captivity is fairly easy for benthic species like Port Jackson Sharks and Small-spotted Catsharks, which are handily caught, transport well, require only small tanks, are not aggressive, and are not finicky eaters. Many such species will even mate in captivity. For these, however, the *Wow!* factor may be missing, since aquarium visitors prefer large, iconic species over those more diminutive and less thrilling.

On the other end of the spectrum are displays of these iconic, larger sharks, which include Whale Sharks, Bull Sharks, Tiger Sharks, hammerheads, Sand Tigers, Sandbar Sharks, Blacktip Sharks, Silvertip Sharks, Blacktip Reef Sharks, and others. All of these sharks must be captured from the wild and transferred to their destination aquariums, often involving long journeys by truck or even plane.

Whether as bycatch, for display, or even in recreational fishing, capturing a shark sets off a suite of potentially deadly physiological responses collectively known as *capture stress*. These include the release of stress hormones and concomitant build-up of chemical by-products of metabolism (e.g., lactic acid), salt and mineral imbalances, hypoxia (low internal oxygen levels), and even dehydration.

The problems that must be overcome to successfully transport a Whale Shark to the Japanese aquarium in Osaka or the US aquarium in Georgia (as much as 8,000 mi or 12,800 km) are daunting, but moving other larger sharks is no stroll in the garden either. All of the large sharks referred to above are ram ventilators, that is, they swim with their mouths slightly ajar, which allows oxygen-rich seawater to flow over their gills. To compensate for the loss of this water flow during transit, oxygen in the form of superfine bubbles may be pushed into the shark's mouth via a submersible pump. Additionally, to allow the heart to perform its role and circulate blood, some flexion of the posterior part of the shark's body is required to return blood back to the heart. Otherwise, blood may pool in the lower and posterior parts of the body, which is unhealthy. Evidence of the latter is reddening of the underside of the shark due to capillaries rupturing.

Once the shark is settled into its final display tank, which may initially involve separating the newly introduced shark with a barrier from veteran sharks already there, the shark will be closely observed and its feeding and nutritional supplementation closely measured.

Even with advances in the process that allows Whale Sharks, the largest fish in the sea, to be displayed, many species simply are not capable of adjusting to long-term captivity. These include the five species of mackerel sharks (White Shark, Shortfin Mako, etc.), which may overheat and succumb to other aspects of capture stress, as well as Blue Sharks and deep-sea sharks, like the Frilled Shark.

Cartilage

Cartilage is the principal structural material of the shark skeleton and it is the essence of what distinguishes sharks from other vertebrates. Bone, which is heavily mineralized and thus harder than cartilage, has found a home in most living vertebrates, about 74,000 species. The *chondr-* of the class Chondrichthyes, which includes the sharks, skates, rays, and the more obscure chimaeras, means *cartilage*.

The cartilaginous skeleton of sharks is tough yet lighter and more flexible than bone. It lacks the nerves and blood vessels found in bone. At stress points like the jaw and vertebrae of sharks, the cartilage is strengthened by incorporating calcium in the form of the mineral apatite. Because its crystalline structure is arranged in a pattern resembling a prism, this type of cartilage is known as *prismatic calcified cartilage*. Chemically, cartilage is composed principally of protein and sugar molecules.

In combination with a buoyancy-enhancing, large, oil-filled liver, the lightness of cartilage is a major benefit, since it helps compensate for the shark's heavy, muscle-bound body and its lack of a swim bladder, a gas-filled internal structure found in most bony fishes that allows them to adjust their buoyancy.

Catsharks and Dogfish

The catsharks are the largest taxonomic group of sharks, with more than 150 species. The term "dogfish" refers generally to the approximately 121 species of sharks in the order Squaliformes, but more precisely to the thirty-seven or so species in the family Squalidae. The catsharks and true dogfish ("true" because the common name of some of the catsharks includes the word "dogfish") are separated by over 200 million years, when the two larger groups to which these belong split apart and diverged evolutionarily. Catsharks and true dogfish have some similarities (generally size and diet, for example) but differ in significant life history characteristics. They are lumped together here, to be honest, because of cliches: By discussing them in one instead of numerous entries, I am killing two birds with one stone. And the phrase "cats and dogs" lends itself to having a *catsharks and dogfish* entry.

Catsharks are mostly small (less than 3.3 ft or 1 m). They occupy cold and deep waters worldwide and are among the 40 percent of sharks that are oviparous (egg layers). The group's common name owes to the resemblance of their elongated eyes to those of domestic cats. They have two small dorsal fins positioned far back on the body.

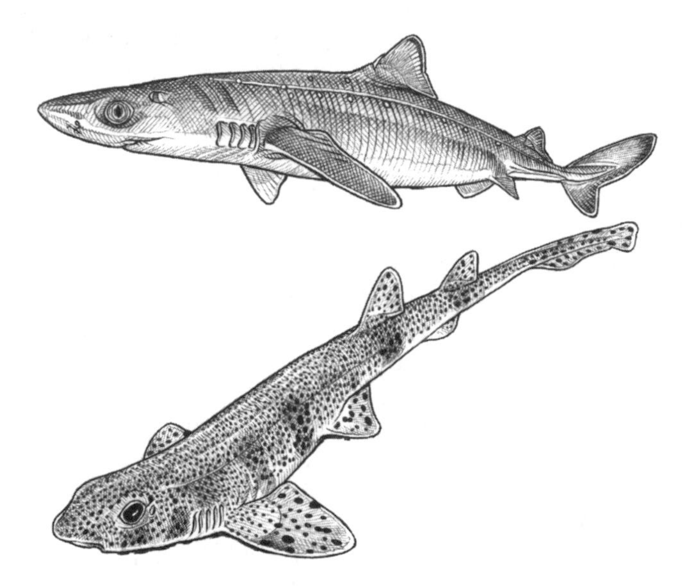

Unless you study deep-sea sharks, you are not likely to cross paths with more than a few catsharks, despite their diversity, and those you see would likely be in marine aquariums or fish markets. One of these is the Small-spotted Catshark, one of the most abundant sharks found in relatively shallow waters in the North Sea, Mediterranean Sea, and nearby. Other shallow-water, more common catsharks include the Chain Dogfish that, despite its name, is not a dogfish, and the Striped Catshark, or Pyjama Shark, found off South Africa and widely displayed in aquariums worldwide.

The true dogfish are found worldwide in tropical, temperate, and boreal seas from the intertidal zone to 1,970 ft (600 m) or greater. A key characteristic of dogfish is the absence of the anal fin, the functional significance of which is unknown. Dogfish are cylindrical in cross section, with two dorsal fins each having strong, ungrooved spines, which may be mildly venomous. A recent discovery of Spiny Dogfish in the Thames River after a long absence prompted sensational headlines of "Venomous Shark Found in Thames." Spiracles on the head are large in dogfish, and the dermal denticles make for tough and abrasive skin.

The best-known dogfish is this same Spiny Dogfish (also called the Spurdog or Piked Dogfish). Despite being overfished in the North Atlantic and Mediterranean, it remains plentiful off New Zealand and is still likely among the most abundant shark species. This species has very conservative life history characteristics (few young, late maturity, and roughly a two-year gestation period). Based on these factors and the depletion documented in stock assessments, the International Union for Conservation of Nature lists it as "vulnerable" globally, with a declining trend. Included here because of its relative abundance and conservation story, the Spiny Dogfish is, in fact, the oddity, since it is one of the few coastal members of the family.

Similarities between dogfish and catsharks include diets consisting of small fish and crustaceans and other invertebrates. Both constitute significant bycatch in deep-sea bottom trawl fisheries as well.

Cephalofoil

Odds are the oddest shark you can envision, and one of nature's animal oddballs, is the hammerhead shark (actually, there are nine known species), whose uniquely

shaped head, or *cephalofoil*, has led to these sharks being called *otherworldly*. Two questions immediately come to mind: How did the head evolve? And what functions does it have?

Evolution favors traits that are on balance adaptive, that is, those that in some way help the organism, for example by saving energy or making it a better predator. One idea is that the widened head serves as scaffolding that can more widely distribute the head's sense organs and lead to enhanced sensory abilities, and indeed that has been found to be the case in terms of electroreception and better binocular vision compared to other sharks. When used as a rudder, the cephalofoil also allows greater maneuverability by narrowing the shark's turning radius. Also, hammerheads are known to use their head to pin stingrays to the seabed while they position themselves to bite.

The cephalofoil ranges in size from that of the relatively small Bonnethead to the bizarrely wide head of the Winghead Shark, an Indonesian species whose head is half as wide as its body length.

Chondrichthyan Tree of Life

The Chondrichthyan Tree of Life project and website represent an ambitious and ongoing venture to document all known chondrichthyan fishes (sharks, skates, rays, and chimaeras), with range maps, illustrations, and CT scans. The information is organized phylogenetically. The evolutionary relationships depicted on the website are estimated based on DNA sequence data for all of the species from which tissue samples could be collected (about 900 of a possible 1,200 described species so far). The project was spearheaded by shark biologist Gavin Naylor and is a collaboration among scientists with different specializations around the world.

A new addition to the shark trait field is Sharkipedia (not to be confused with the similar title of this book). It advertises itself as an "initiative to make all published biological traits and population trends on sharks, rays, and chimaeras accessible to everyone." Science works better with community-minded contributions like these two examples.

Claspers

All sharks (and skates, rays, and chimaeras) use internal fertilization, as opposed to the female depositing unfertilized eggs in the environment. Transferring the sperm internally while swimming is tricky, which explains in part why internal fertilization is uncommon among the bony fishes. Additionally, females may be unreceptive to the male's attempt at mating, making you wonder

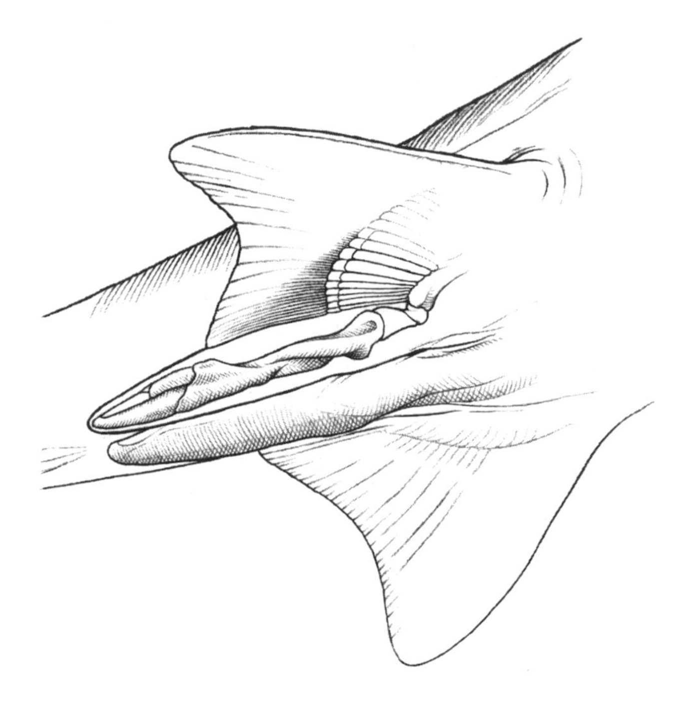

how internal fertilization in sharks survived natural selection's rigorous standards.

The answer to this evolutionary challenge is *claspers*, structures that Aristotle himself observed and named, although he was mistaken in his notion of how they functioned. Also called *mixopterygia*, claspers are rearward tubular modifications of the inner margins of the pelvic fins of male sharks (and a distinguishing feature of chondrichthyan fishes). Aristotle erroneously surmised that the claspers were employed by males to grasp the female while he fertilized eggs externally.

Although the explanation was wrong, the name has endured.

During fertilization, a single clasper is inserted into the cloaca (the common urinary, genital, and anal opening) of the female, after which insemination occurs.

Why two claspers when only one is used at a time? The clasper that is used is rotated ninety degrees or more across the body before insertion. Having two claspers enables access to the female from either side, which comes in handy if multiple males are competing to inseminate the female.

The actual sperm transfer occurs via a groove in the claspers. After the clasper splays open and anchors itself in the cloaca with hooks or spines, or simply enlarges sufficiently to remain in the cloaca, sperm are flushed into the female along with a relatively large volume of seawater that the male had previously imbibed in a specialized siphon sac. After the act, the clasper is disgorged, which may require forceful shaking by the female.

Coastal Sharks

Of all shark habitats, coastal waters are among the most biodiverse. At the same time, because nearly 60 percent of the human population live near the coast, these areas are heavily degraded by pollution and habitat alteration.

Nearly 40 percent of shark species live in the coastal waters of temperate and tropical areas. Many of these areas are characterized by abundant food resources and very high habitat diversity. Coastal habitats include estuaries, kelp forests, seagrass beds, live bottom habitat, and beaches. Here, we focus only on temperate coastal waters (tropical and polar sharks are considered separately).

Estuaries, where freshwater and saltwater meet and intermix, are often found at the mouths of rivers. Today's estuaries are all less than twelve thousand or so years old, resulting from the geologically recent rise of sea level since the last glacial maximum. Their shallow, nutrient-rich waters, along with the intertidal wetlands that line their shores, are some of the most productive ecosystems on Earth. About fifty species of sharks, ranging from neonates (newborns) to adult apex predatory sharks (e.g., Bull Sharks), are found in estuaries. Other sharks commonly found in estuaries include Sandbar Sharks, Lemon Sharks, Blacktip Sharks, Blacknose Sharks, Atlantic Sharpnose Sharks, Pigeye Sharks, Port Jackson Sharks, and Bonnetheads.

In larger estuaries, different shark species may occupy different parts of the system. In Winyah Bay in northeastern South Carolina, one of the largest estuaries on the Atlantic Coast, as many as ten species of sharks in an array of sizes and ages are commonly found in the warmer months. Larger species (greater than 5.7 ft or 1.75 m) include Lemon Sharks, Bull Sharks, Sandbar Sharks, Finetooth Sharks, Spinner Sharks, and Blacktip Sharks. Adults of smaller species, including Atlantic Sharpnose Sharks, Blacknose Sharks, and Bonnetheads, also inhabit the system. Juvenile Sandbar Sharks take advantage of their ability to tolerate the lower salinities, an adaptation that most adult sharks (even Sandbar Sharks) lack, to occupy the middle bay region, where it is much less salty. This juvenile adaptation—tolerance of low salinities—allows them to use the area as a refuge from the predation of most larger sharks.

Of the remaining coastal habitats, the harshest is the zone nearest the beach. Beaches are high-energy, dynamic, unstable environments with coarse sediments, a perfect storm of harsh factors that limit the abundance of both sharks and their prey. Schools of small fish, known as *bait balls*, will attract Blacktip Sharks, Sandbar Sharks, and others near the beach, often creating exciting theatrical displays of huge splashes and charging shark bodies piercing the surface in pursuit of a meal.

Kelp forests, large offshore areas of fast-growing algae that flourish in clear, primarily shallow water, are very productive environments that provide important habitat for a variety of sharks. These include, in different locales, Horn Sharks, Leopard Sharks, Swell Sharks, Pyjama Sharks, Happy Eddie Shysharks, and Broadnose Sevengill Sharks.

Sharks found in other temperate-water coastal ecosystems—for example, on live bottoms—include Spiny Dogfish, Pacific Spiny Dogfish, Leopard Sharks, Dusky Smoothhounds, Tiger Sharks, hammerheads, and a large number of requiem sharks.

Common and Scientific Names

Where I grew up, on the coast of the Southeastern United States, I thought there were only three kinds of sharks: "sand sharks," Blacktips, and hammerheads. To this day, many residents and visitors to the area think similarly. In actuality, the number is two dozen or more, depending on the area. And herein lies the danger of using common names: because the same common name is frequently used for several species, they deny an organism—with its own distinct genes, life history, behavior, and distribution—its earned unique identity, and the consequences from this could be endangering.

In science, *species* refers to organisms that freely breed with each other and that produce viable offspring capable of producing their own offspring at some point. There are currently about 541 distinct species of sharks, although more are likely to be discovered using modern molecular techniques, or as the deep sea is better explored. Each species has a unique two-part scientific name. Almost every species also has one or more common names, even in the same locality. The hammerhead sharks of my youth could have been one of five different species. But we considered them all one species—hammerheads—and blissfully ignored the biodiversity.

Recently, common names for a large number of sharks have been standardized. The accepted common names of the hammerheads of my youth are Bonnethead, Scalloped Hammerhead, Smooth Hammerhead, Great Hammerhead, and Carolina Hammerhead. Note that the first letters are upper case, which is not yet a universal convention—although it should be! (Why do we capitalize the first letters of corporations but disrespect our sharks, trees, and so on?) Also note that, for these sharks, the word "Shark" is not a part of the common name, nor is it for Shortfin Makos, Sand Tigers, and a few others.

How can using common names endanger sharks? Spiny Dogfish (*Squalus acanthias*) and Dusky Smoothhounds (*Mustelus canis*), discussed earlier in the entry for "Catsharks and Dogfish," overlap in their distribution along the US Atlantic Coast and, to many, resemble each other. Prior to 2002, both were heavily fished and were caught, occasionally in mixed schools, on the same fishing gear. For fishery statistics, which are invaluable to fishery managers and conservation biologists in regulating any fishery, both species were considered in a single category, *dogfish*.

Lest you think classifying them together was innocuous, consider this: The two species are in different superorders (Spiny Dogfish are squalomorphs, and Dusky Smoothhounds galeomorphs), groups that diverged about 210 million years ago. More significantly, their life history characteristics are drastically different. It takes Spiny Dogfish as long as 20 years to mature, after which they give birth to only 4–6 pups every other year, after a two-year (!) gestation period. Smooth Dogfish, aka Dusky Smoothhounds, are at the other extreme. They mature in only three years and have an average of 12 pups annually after a gestation period of 8–9 months. The northwest Atlantic stock of the former species plummeted before fishing was restricted and after a period of rebuilding, mature females and pups have declined again from overfishing. Smooth dogfish also have catch limits and the fishery appears sustainable for now.

One final point: *Carcharodon carcharias* is the White Shark, not the *Great* White Shark. We all know its greatness without that reminder!

Convergent Evolution

Organisms that are not closely related, or perhaps are even in different taxonomic phyla, sometimes have similar anatomical, behavioral, or physiological adaptations. In some of these cases, where ancestors of the different groups lacked the adaptation, these similarities are evidence of *convergent evolution*. Convergent evolution can be thought of as distantly related organisms enhancing their survival in analogous ways. Traits that are the result of convergent evolution are fascinating and they foster a deeper understanding of the traits involved. At the same time, these traits are of little use in understanding an organism's evolutionary history and current relationships. To understand these issues, *homologous* traits—those that result from common ancestry—are required.

There are copious examples of convergent evolution in nature, the most recognizable being the presence of wings among diverse animals such as birds, bees, flying fish, bats, and so on.

The most superlative example of convergent evolution in fishes, including sharks, is the suite of adaptations associated with high-performance swimming and superior predatory abilities among three lineages: the lamnid sharks (e.g., White Shark and Shortfin Mako), the thresher sharks, and about fifteen species of scombrid fishes (tunas). These sharks and tunas are in two taxonomic classes that diverged as long ago as 450 million years.

While studying Shortfin Makos on the Pacific Ocean, I snapped a photo from above of an Albacore tuna and a Shortfin Mako posed next to each other. When the photo was developed (no digital photography then), I was stunned by the similarity of form. Both were heavily muscled, nearly perfectly streamlined oceanic projectiles with superficially similar tails and a broad, flattened, keeled posterior.

The pièce de résistance, though, is the evolution of their regional endothermy, maintaining their internal temperatures above that of their environment, a rare feat among aquatic organisms because water holds a lot of heat and removes it quickly from warmer objects, making it a challenge to be warmer than the water.

The advantages of endothermy are extreme swimming performance that enables endotherms to be highly mobile and efficient predators, and possibly to move more independently of environmental temperature than other fishes. There is also a cost: feeding the metabolic beast requires more food. But being exceptional predators takes care of that problem!

Cookiecutter Shark

Imagine you are a large, formidable, bony fish or marine mammal (or even a nuclear submarine, see below), fearlessly moving through the ocean's surface waters in the evening. Threats to your existence could come from a bigger or more ferocious predator, but none are present, so you relax and let your guard down. And what luck—lurking just above you is an unsuspecting small fish, a tasty, easy prey for such an impressive predator as you.

Bad move! The small fish is, in fact, subterfuge, a lure to attract you, and you took the bait. It is no fish at all, but actually a darkly pigmented body part that resembles the silhouette of a small fish from below, and its owner is a cigar of a shark, with a disproportionately sized set of razor-sharp, triangular lower jaw incisors and an otherworldly face, at the boundary between silly and diabolical. Were it the subject of a horror film, this shark would be the stuff of nightmares.

When you are close enough to smugly anticipate the ease with which you will catch this poor fish, it engages its supercharger and jets to you, using its powerful caudal fin, to remove a plug of your musculature with a death spiral that operates at near surgical precision.

Okay, so I took some liberties with anthropomorphism and hyperbole. But there is such a shark, the Cookiecutter Shark. They are neutrally buoyant vertical migrators, and they have luminous organs that emit light that matches the ambient downwelling light, so the silhouette of the shark disappears from beneath, except for the collar on the underside—the bait that fools many an erstwhile predator.

Cookiecutter Sharks are small (up to 2 ft or 0.6 m). About those nuclear submarines? Yes, Cookiecutter Shark scars have been seen on some rubber parts of these. There are also cases of attacks on people over deep water at night off Hawaii. If Cookiecutter Sharks were common along the shore, I'd rethink evening dips!

Countershading

The name of the game in evolution is survival, and paramount to survival is finding your food and avoiding being someone else's. One common strategy in this game of life employed by sharks and other animals is stealth, and a widely used stealth method is countershading, in which the upper surface is dark and the underside light.

Viewed from below, a countershaded shark blends in with downwelling light. From above, it matches the darkness of the depths.

Numerous sharks (and other species) of the well-lit surface layer are countershaded, as are others in the deeper oceanic twilight zone. Coloration may be from skin pigmentation or bioluminescence. In both cases, a countershaded shark is less conspicuous to both predators (typically bigger sharks) and prey.

Countershading on the Shortfin Mako, with blue on top and white on the underside, is striking. Why would such a high-performance shark, capable of outswimming both predators and prey, require countershading? First, Shortfin Makos, especially juveniles, do indeed have predators, mostly larger sharks. Second, as prodigious as they themselves are as predators, any advantages that make them more successful and save energy in the process will be favored by evolution.

In addition to countershading, sharks employ other strategies, including camouflage, or cryptic coloration. Neonate Nurse Sharks, for example, candidates for the title of *most adorable shark* in anyone's book, have bars and spots that enable them to blend into their shallow water benthic habitats, and thus avoid predators who consider them more tasty than adorable.

CSI: Shark

Among the less attractive activities I have been asked to do is confirming that a wound, typically on a person's arm or leg, is in fact a shark bite. Along the northeast coastline of South Carolina, there are typically several shark bites during the summer, when humans and fast-swimming Blacktip Sharks, thought to be the culprits, coexist in the warm, murky beach shallows. Somehow the news media obtains gruesome photos of some of the wounds and, as the only nearby shark specialist, I invariably find these snaps in my inbox. Inasmuch as shark bites are not my specialty, I limit my assessment to either *yes, it looks like the bite of a shark*, or *no, you can't blame a shark for that one*. Mostly it is the former.

In another book about sharks that I coauthored in 2020, *Shark Biology and Conservation*, I relate a story about an actress in the adult film business who falsely claimed that a shark bit her during a dive. Conveniently for her, the entire episode was filmed—sharks in the water, screams, and then the actress emerging with a bloody but clean laceration. If one of the large sharks in the video had actually bitten her, it would have either removed a sizeable chunk of flesh or left a series of ragged-edge tooth punctures in an arc corresponding to the arched shape of a shark jaw—or both—but it would not have resulted in a clean, straight laceration. This "bite" was most likely self-inflicted with a sharp (but not shark) implement in the opinions of most shark biologists.

What would a trained analyst of shark bites look for, besides puncture marks in an arc? First, the bite radius can give an indication of the size of the shark. Second, the shape of the arc can be diagnostic (e.g., broad or narrow). Third, because teeth can vary among shark species, as well as between the upper and lower jaws of some species, any impressions left by the puncture marks and the spacing between can also be invaluable. Even under the best of situations, because there is no database of shark bite forensics, identifying the kind of shark responsible usually is at best an educated guess, the accuracy of which depends on the investigator's background, experience, and knowledge of local sharks. Having a museum nearby with a collection of shark jaws from local species can help as well.

Shark bite forensics is of interest beyond human bites. Sharks have been known to bite underwater cables and sonar arrays towed behind submarines. Knowing which shark was the culprit in these cases can help to deter the perpetrator from future interactions or mitigate the damage.

Daily Ration

How much does a shark eat in a day? As you might suspect, the answer varies with the species and size. And what you are really asking is, *What is a shark's daily ration?* Since species vary in their weight generally and in different life stages, daily ration is typically reported as the mean percentage of an organism's total weight consumed over a twenty-four-hour period.

Knowing an organism's daily ration is more than merely trivial information. Along with life history characteristics, understanding a shark's ecological role and metabolic needs (amount and kinds of food it must eat to do what a shark does, e.g., swim, breathe, etc.) are critical to determining its conservation status and managing it.

On average, sharks consume between 2 and 3 percent of their body weight per day, a range that varies depending on the energy demands of a specific shark species or life stage and the energy content of the prey.

You might expect that an adult Shortfin Mako, the bravura Bugatti of blue water, might have a significantly higher daily ration than less active sharks because of the Mako's need to feed its calorie-hungry metabolic machinery. The data, however, are equivocal, with estimates ranging from 2.2–3.0 percent in one study to greater than 4 percent in another. For an average-sized Shortfin Mako of the Northeastern United States, this translates into about 1,100 lb (500 kg) of bluefish during their half-year residency.

If the unexpected lower estimates are real, the Shortfin Mako's calculated daily ration could reflect higher digestive efficiencies or more calorically dense prey. Alternatively, the data are inexact, since studying the Shortfin Mako is fraught with logistical challenges, as stated in the entry for "Elasmotunatron."

Deep-Sea Sharks

Imagine if nearly half of the planet's bird life lived at altitudes so high that eyeing one was a rare event, and birds thus remained mysteries to the public and scientific communities. Such is the case with deep-sea sharks. That very proportion of shark species inhabit the deep sea, which is generally defined as depths below 660 ft (200 m).

Our lack of scientific understanding of deep-sea sharks is both logistical and economic. Large ships are expensive to operate, costing up to tens of thousands of dollars a day. While submersibles are in wide use in deep-sea industrial applications, only a handful are available for scientific exploration and, you guessed it, these are expensive as well. That leaves the last resort as

commercial fishers, not all of whom are willing or able to share their data.

It should not be surprising that such a large diversity of sharks occupies the deep sea, since it is the largest ecosystem on the planet. In the mesopelagic zone, between 660 ft (200 m) and 3,300 ft (1,000 m), sharks are the dominant predators.

Abundant living space, however, does not necessarily equate to hospitable conditions. The deep sea is cold, dark, and the hydrostatic pressure is high. At 6,500 ft (2,000 m) deep, the pressure is not unlike the entire weight of a moose balancing on your nose. Finally, major sections of the deep sea are food-poor environments. Because the environmental features of the deep sea (pressure, temperature, salinity, and light levels) are similar over wide vertical and horizontal expanses, many shark inhabitants have very broad, sometimes global, distributions.

Living in the cold, dark, pressurized, food-poor deep sea environment means evolving adaptations to conserve energy, maintain internal function under extreme conditions, locate prey, avoid being prey yourself, and find mates. These adaptations translate into physical, physiological, and behavioral departures from the fast-swimming, large-bodied, gray coastal shark cousins; deep-sea sharks look different and, in some cases, un-sharklike.

Adaptations among deep-sea sharks to their unique suite of environmental challenges include specialized eyes, photophores (for bioluminescence), lower metabolic rates and activity levels, varieties of teeth that ensure that prey, once captured, do not escape, and year-round breeding.

Sharks of the mesopelagic zone include Cookiecutter Sharks, Goblin Sharks, Frilled Sharks, Gulper Sharks, Dwarf Lantern Sharks, and Cuban Dogfish. Beneath this zone are Portuguese Dogfish and Bluntnose Sixgill Sharks. The depth record for sharks is held by the Portuguese Dogfish, about 12,000 ft (3,700 m), which is noteworthy because at depths below approximately 9,800 ft (3,000 m), the oceans are almost completely devoid of sharks. Explanations for this absence include limited food resources and the inability to synthesize important chemical compounds under the high ambient pressures.

Dermal Denticles

Like bony fishes, the bodies of sharks are covered with scales, specifically *placoid* scales, or *dermal denticles*. These are essentially miniature teeth, with an inner pulp cavity, surrounding layer of dentin (hard, calcified tissue), and a hardened outer layer of enamel. Like their teeth, dermal denticles are shed and replaced, although more slowly.

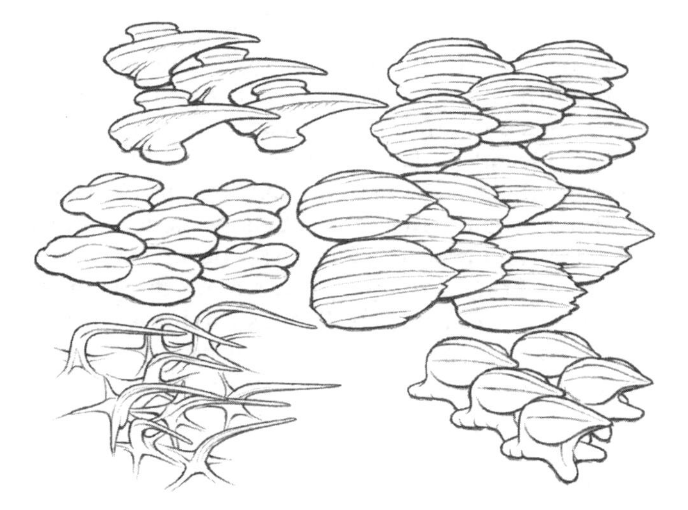

The scales of bony fishes differ in that they contain mineralized bone and are permanent.

Scales have different functions among sharks, including reducing drag while swimming, protecting against ectoparasites, and safeguarding females from male bites during mating season. Scales vary in size, shape, flexibility, and even coverage in sharks. Bramble Sharks, for example—large, sluggish, poorly known deepwater sharks—have large scales scattered over their body. Spines on the fins of sharks and the whiplike tail of stingrays are modified dermal denticles that are primarily defensive in nature, although in sharks they may have a hydrodynamic function as stabilizers.

The drag-reducing function of the dermal denticles was the inspiration for a full body swimsuit, the Fastskin FSII, developed by Speedo in 2014. Reducing drag, the major force that slows moving objects, is the name of the game in swimming for both sharks and elite competitive swimmers. The Fastskin FSII was marketed as a revolutionary, performance-enhancing body covering that significantly reduced drag. According to Speedo, the fabric channeled water smoothly from front to back along the swimmer's body, in much the same way that the ridges and grooves of a shark's dermal denticles were thought to do. Although the suit produced faster swimming speeds, humans do not swim like sharks, so further investigation of the fabric's mechanism of action is warranted.

Diet

The public perceives that sharks are indiscriminate eaters with insatiable appetites. There is some truth to the former, at least among some species that are opportunistic or generalist predators, but the latter is myth.

Selection of prey often varies on multiple time and space scales. Most sharks are generalists with diverse diets. When prey abundance or choice changes, they can switch prey types. Numerous studies support the conclusion that sharks will consume the most abundant prey available. When we longline for sharks for research, education, and conservation, at times of year when small baitfish like mullet, menhaden, or spot are very abundant, we often catch few or no sharks, when at other times we'd catch ten or more. They forage on the baitfish more so than our bait. Wouldn't you prefer fresh to frozen?

Some shark species specialize in certain prey types. For example, Horn Sharks feed primarily on hard-shelled mollusks and crustaceans, Dusky Smoothhounds prefer crabs (especially recently molted crabs—talk about finicky eaters!), Frilled Sharks eat mostly squid, and Bonnetheads selectively devour Blue Crabs.

How do scientists know what sharks eat? Two approaches are most often used: analysis of gut contents and stable chemical isotope ratios. In the former, which is still the most robust method for studying a shark's diet, the stomach contents of dead sharks are removed in the field, preserved or frozen, and then identified in the lab, a meticulous process that might require identifying a species of fish by the presence of a few bone fragments. Flushing the gut contents from the digestive track of living sharks with water, called *stomach lavage*, is also practiced, followed by release of the live shark. *Don't try this at home!*

The second method, analysis of stable isotopes, assumes that *you are what you eat*. All that is needed is a small muscle plug, blood sample, or piece of skin, after which the shark can be released. Analysis of these isotopes can provide information on the shark's trophic level and whether the shark is feeding in a benthic or pelagic food web. A major weakness, however, is not identifying the actual prey species.

Finally, you can infer much about the diet of a shark from the morphology of its jaws and teeth. Sharks like Blacktips and Sandbars have narrow, cusped lower teeth for grasping prey, whereas the upper teeth are slightly wider with lateral, sharper edges that allow them to slice prey into pieces. Bull Sharks have triangular upper teeth for cutting bigger chunks from larger prey. The horn sharks have cusped teeth up front, molariform (flattened) teeth in the back, and hypertrophied jaw muscles for crushing snails, urchins, and crabs.

Many sharks switch their prey as they grow (this is known as *ontogenetic diet shifts*). The Shortfin Mako swallows some of its prey whole (bony fishes and cephalopods, predominantly), but as they age, their teeth become broader and flatter, enabling them to widen their prey options to include organisms too large to swallow whole but from which they can remove a chunk of flesh (e.g., Swordfish, tuna, sharks, sea turtles, and marine mammals).

Finally, no species of shark includes humans as regular menu items, but you knew that already.

Diversity of Sharks

This book has repeatedly referred to the 541 known species of sharks. If we include their close cousins, the batoids and chimaeras, the number rises to around 1,300, or 1.7 percent of known living vertebrates. However, shark taxonomists do not honestly know how many species there are. The number is increasing due to modern molecular techniques that distinguish between species closely resembling each other, as well as increased sampling in the deep sea and remote coastal regions.

Sharks have indeed been successful, but before we shark enthusiasts become smug about their success, consider that the dominant aquatic vertebrates are the approximately 38,000 kinds of bony fishes. Species of catfish are even more numerous: there are about four thousand, which is more than sharks, batoids, and chimaeras combined.

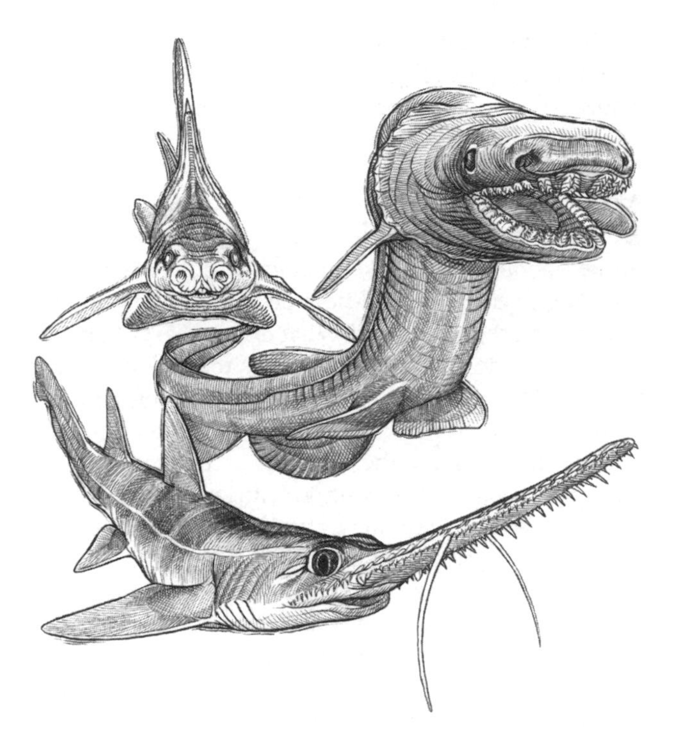

The diversity disparity between sharks and bony fishes has several explanations. Most concern increased *vagility*, or dispersal, one of the major preconditions for some types of speciation. For example, bony fishes have more offspring than sharks (orders of magnitude more, in some cases), and greater numbers equals greater dispersal. Most adult sharks are small, less than 3.3 ft (1 m), and most of these are bottom-associated. As a rule, small, benthic sharks don't disperse much and thus do not form new species.

There may be fewer sharks than bony fishes, but that in no way diminishes the incredible array of fascinating shapes, sizes, colorations, behaviors, and so on, of the sharks, or our appreciation of them!

Ecology/Ecological Roles

Shark ecology deals with some big questions about how sharks interact with their living and nonliving environment. Where do sharks live? What are their habitat requirements? What do they eat? What ecological roles do they play? Answers to these questions are critical to understanding the ecological roles and life history characteristics of sharks in order to effectively manage and conserve them as well as the ecosystems in which they live—not to mention sating our own curiosity.

Sharks inhabit habitats as varied as the open ocean, deep sea, estuaries, and coastlines of virtually all latitudes, and even rivers. Where sharks are found depends first on the environmental characteristics—temperature and salinity are of paramount importance—of that location and the ability of the shark to adapt to them. There are also ecological constraints to consider, for example, the presence of suitable prey at appropriate densities, or the competition from other sharks and other species for that prey. For smaller species of sharks or juveniles, the risk of predation must also be weighed. Thus, successfully occupying a particular habitat requires physiological tolerance of the nonliving environment and a suitable balance between competition for resources and risk of predation.

Determining the roles that sharks play in their ecosystems is a surprisingly difficult ecological question. Sharks as a group are often mistakenly characterized as apex predators, that is, at the top of the food chain and having few or no predators themselves. For a few shark species, of course, this is indeed the case. But most sharks, in fact, are mesopredators (mid-level predators), one or more trophic (feeding) levels beneath the apex predators. Mesopredatory sharks thus have their own predators, and these may include birds, crocodilians, marine mammals, large bony fishes, and even other sharks.

Apex predators are thought to maintain ecosystem health by controlling prey populations and eating the weak and diseased, but this is an oversimplification for which supporting empirical evidence in the case of sharks is simply not available. Mesopredators also play important roles as both predator and prey in their ecosystems.

On coral reefs, Tiger Sharks, Bull Sharks, Great Hammerheads, and Silvertip Sharks are considered true apex predators. Elsewhere, others include White Sharks, Oceanic Whitetip Sharks, Greenland Sharks, and Bluntnose Sixgill Sharks. In mangrove estuaries, Bull Sharks and Smalltooth Sawfish (a batoid) share the top predator space. Other species, such as Caribbean Reef Sharks, may be part-time apex predators in their coral reef habitat in the absence of larger apex predators.

It's tough being an apex predator, since they face unique challenges because they begin with a huge disadvantage: there are fewer of them, a fact of life determined in part by the laws of thermodynamics (less energy is available at every feeding level as you move up the food web, and thus less food to support the top levels).

Sharks play many roles in marine ecosystems that are often complex and not well-known. Much more work must be done for a more complete understanding of the ecology of the 541 species of sharks in their varied environments, especially as human impacts accrue.

Ecotourism

How much is a shark worth? The answer depends on whether you are a commercial fisher, fish monger, fin distributor, scuba diver, conservationist, philosopher, and so on. Although it is a perfectly valid idea to conclude that not everything in creation can or should have a dollar value attached, economists have made such calculations. In The Bahamas, an estimate of the lifetime economic value of a Caribbean Reef Shark was approximately $250,000 USD as a tourism draw, but only around $50 USD if killed.

Shark ecotourism is growing. A 2011 paper identified 376 shark ecotourism operations. By 2013, according to another study, there were nearly 600,000 people who traveled for shark ecotourism annually. Whale Sharks, White Sharks, Tiger Sharks, Great Hammerheads, and Caribbean Reef Sharks—a veritable who's who of iconic species—top the list of sharks in ecotourism, along with Manta Rays. As such species represent a small subset of sharks and rays, using ecotourism value as a conservation driver is a seriously limited approach.

The benefits of shark ecotourism are indisputable, and include bolstering local economies, and, in the process, serving as incentives to displace practices that are environmentally harmful, like overfishing or habitat degradation. Seeing sharks in their habitat peacefully doing what sharks do, including eating, supplants the horrific pictures imprinted in our heads from the violent depictions that bombard us in the media, and this can play an important role in countering negative perceptions.

The benefits come with costs, and these potentially counteract some of the positives. Consider provisioning—that is, attracting sharks to baited stations—and even hand-feeding them. Some sharks at these stations may be provided with enough food to meet their entire metabolic needs for a day or more. However, the food may be nutritionally insufficient. Moreover, provisioning may impact predator-prey dynamics in their habitat and change habitat use, attracting sharks beyond their natural home ranges. It may modify the timing and pathway of migrations, alter the behavior of the sharks, and habituate them to humans and associating people with food, which may pose public safety risks.

Once an area develops its shark ecotourism, it may become so dependent on it that it suffers greatly if sharks leave the area or decrease in number. False Bay and Gansbaai, South Africa, are historically two of the most reliable sites for spectacular White Shark sightings. In 2017 and 2018, White Sharks disappeared from these locations for long periods, leaving the diving operations as well as the support services in danger of financial ruin. The cause of the sharks' absence is not yet clear, but plausible explanations include recent visits by Orcas, which killed and eviscerated several White Sharks, as well as poaching, and climate change. Whether this is a short-term phenomenon or one that will endure is not known.

Finally, underappreciated but nonetheless significant negative effects of ecotourism include the environmental and social impacts of travel, for example, emissions, resource use, consumerism, demands placed on local utilities, higher costs, and so on. Also, subsistence and artisan fishers may lose access to valuable fishing sites.

Elasmobranch

The word *elasmobranch*, which is pronounced *ee-lăz'mo-brank*, translates from its Greek roots roughly to "plate gills," a reference to the interbranchial septa (partitions) to which the gills are attached and that resemble stacked plates. Another interpretation is that the word means *strapped gills*, because the gills appear to be strapped together by the septa.

Taxonomically, elasmobranchs are in the subclass Elasmobranchii, which includes the sharks, skates, and rays. Together with its sister group (the closest related taxonomic group), the subclass Holocephali, or the chimaeras, they form the class Chondrichthyes. Elasmobranchs possess dermal denticles, protrusible jaws of varying extents, serially replaced teeth, and pectoral and pelvic fins supported by three-part cartilage. Their gills lack the level of protection afforded by the operculum of bony fishes and are sometimes referred to as *naked.*

Elasmotunatron

Among the areas of interest for physiological ecologists, scientists who study how internal function relates to an organism's ecology, is an organism's metabolic rate, or energy use over time. Knowing this helps determine an organism's energy budget, an account of all the energy required by an animal and how it is expended. This information has practical value, for example, in managing fisheries. As overfishing and global climate change impact food supplies, activity levels, and other physiological and behavioral processes, it becomes critical to understand the energetics of sharks in their ecosystems.

The method most commonly used to determine an animal's metabolic rate is *respirometry*, in which sharks swim in circles or against a current in a tank. Respirometry does not measure the actual use of energy but rather oxygen consumption, which correlates with (that is, serves as a proxy for) energy use. Studies on smaller sharks have demonstrated that metabolic rate correlates with activity level, and these rates are similar to those of bony fishes with similar ecological roles.

But a big gap in knowledge exists for species like the Shortfin Mako, a streamlined, swift-swimming super-predator. The logistical problems of capturing a healthy, unstressed Shortfin Mako make bringing specimens back to the lab impossible. Enter Scripps Institution of Oceanography physiological ecologist Dr. Jeff Graham and colleagues (I played a minor role as a doctoral student in Jeff's lab), who, in 1982, codesigned and supervised the construction of one of the biggest treadmills on the planet, a *water* treadmill called the *elasmotunatron*. This behemoth was constructed to determine the metabolic rates of swimming Shortfin Makos and Albacore tuna at sea from the deck of a ship. The major finding from swimming nine Shortfin Makos in the elasmotunatron was that the maximum metabolic rates were among the highest measured for any shark species, results consistent with the Shortfin Mako's reputation as a predator of incredible performance levels.

Endangered Sharks

Ignoring for the moment the fact that all of the biosphere is endangered by imminent, large-scale threats, are there any sharks on the verge of critical population decline that might put them on a trajectory to extinction? Unfortunately, indeed there are.

Globally, the International Union for Conservation of Nature (IUCN) Shark Specialist Group assesses the risk of extinction for sharks and monitors threats. In the United States, the 1973 Endangered Species Act governs the process. As of 2021, about one-third of sharks, skates, rays, and chimaeras are listed by the IUCN as "threatened," a catchall category that includes "critically endangered," "endangered," or "vulnerable" (*endangered* denotes a *very* high risk of extinction; *critically endangered* means a species is considered to be facing an *extremely* high risk of extinction). The biggest current and historical threat has been fisheries, both as targeted catch and bycatch. Without substantial additional limits on fishing, the fraction of species deemed threatened will undoubtedly rise.

Extinction risk is greatest for warm-water, coastal sharks and rays; more than three-quarters of tropical and subtropical species are threatened. Three species not reported for many decades (the Lost Shark, Red Sea Torpedo Ray, and Java Stingaree) are now considered possibly extinct. The most threatened chondrichthyan families include the sawfishes, wedgefishes, giant guitarfishes, devil rays, and pelagic eagle rays, all of which are batoids, as well as hammerheads and angel sharks. Overfishing is the main risk for all of the 391 threatened chondrichthyans.

A comprehensive listing of threatened sharks can be found on the IUCN Red List website. The following sharks are classified by the IUCN as "critically endangered," meaning that they are facing an extremely high risk of extinction: Pondicherry Shark, Ganges Shark, Daggernose Shark, Striped Smoothhound, Oceanic Whitetip Shark, and several species of angel sharks, wedgefishes, and sawfishes.

The IUCN Red List is not to be confused with species listed under the US Endangered Species Act (ESA). The Red List serves to highlight at-risk species on a global scale but carries no mandates for protections or policies. Different criteria are used to determine listing (as "threatened" or "endangered") under the US Endangered Species Act. The first elasmobranch listed as endangered under the ESA, in 2003, was the Smalltooth Sawfish. Strict prohibitions on a range of harmful activities were enacted as a result. Since then, a comprehensive recovery plan has been developed and critical habitat (which triggers agency consultation for development projects) has been designated. In 2014, two distinct populations of the Scalloped Hammerhead, those in the Eastern Pacific and Eastern Atlantic Oceans, were listed as endangered, but there is currently no recovery plan. In 2018, the Oceanic Whitetip Shark was listed as threatened. The Giant Manta Ray was also listed as threatened under the ESA in 2018. Other ESA-listed elasmobranchs that have exclusively foreign distributions include the four additional sawfishes, five angel sharks, three guitarfishes, two smoothhounds, and the Daggernose Shark.

Science-based management has protected some elasmobranch species and helped others onto a trajectory to recover. However, over the last thirty years, these approaches have subordinated sharks (and, to a greater extent, rays) in favor of more economically important groups, such as tunas, Swordfish, flounder, and cod.

If you ever encounter an endangered elasmobranch in the wild—for example, a sawfish—you should not approach it or attract it, but you should report it to authorities. The US has a shared database for Smalltooth Sawfish encounters. You can report a sighting at 1-844-4SAWFISH or by emailing info@sawfishrecovery.org.

Fear of Sharks

Fear of Sharks is known as either *galeophobia* or *selachophobia*. I never feared sharks until I was surrounded by large, actively feeding Caribbean Reef Sharks while snorkeling in the warm azure waters of Bimini in The Bahamas in the 1990s. A few minutes prior, several slender, unintimidating Caribbean Sharpnose Sharks had been attracted to the nearby dive boat. Easy-peasy! That is, until a posse of stout Caribbean Reef Sharks made their presence known, at which my body reflexively resorted to what every vertebrate does when confronted with potential danger: it engaged its sympathetic nervous system with the resulting *fight or flight* response—that is, increased alertness, racing heart, faster breathing. What I did not do, fortunately, was panic and, after a few calming breaths, when it was clear that these amazing husky beasts were not targeting me, I relaxed just enough to let myself be immersed in awe.

Our fear of sharks, and other predators as well, is likely both innate and learned, and had survival value in a world long ago. For much earlier generations, sharks and other predators posed a real danger, and evolution has endowed us with an appropriate rapid protective reaction.

But most readers are under no immediate threat from predators now. Our innate response, however, remains. It is also made worse by the fact that we are assaulted by negative shark stereotypes in the media, and especially television. These stereotypes sometimes worm their way into our psyches and overwhelm our rational thoughts; many who know that a shark bite constitutes a low risk still experience fright disproportionate to the real threat.

Humans also fear the dark, spiders, snakes, heights, public speaking, a visit to the dentist, returning a library book late (well, I do), and so on. Our brains conjure up reasons: What evil could be lurking in the dark? What if I fell from this scenic overlook? What if I stepped on a Gaboon Viper that I did not see? What if the librarian embarrasses me? And the more we contemplate these unlikely occurrences, the more we nurture our fear. Yes, fear evolved in humans as a survival mechanism, and there are times when it is rational to be fearful, but much of our fear borders on panic and is not based on the high likelihood of a harmful thing's actual occurrence.

Fear of sharks is most like fear of the dark, which is really fear of the unknown. For all a typical beach-goer knows, there could be hundreds of starving sharks salivating in anticipation of their entry into the surf, especially in murky water. And layered on top of this fear are the gruesome descriptions and images of shark attack victims. The fact that a shark attack would be a particularly horrifying way to die, or to lose a limb, is the icing on the cake when it comes to fearing sharks.

Want to overcome your fear of sharks? Observe and learn about them, and then possibly swim with them, with trained supervision. And replace galeophobia with galeo*philia* (love of sharks).

Feeding Frenzy

During my annual Biology of Sharks course at the Bimini Biological Field Station, we snorkel with actively feeding Caribbean Reef Sharks. A dozen or more sharks may gather near the boat, awaiting a chunk of barracuda that will be tossed into their midst, while students in the water

observe from a safe distance. At the sound of the splash, the sharks race toward the food and the swiftest typically wins the prize. When a handful of food is tossed, the scene becomes slightly more hectic, but the food is quickly consumed, and the sharks retreat to await more food.

Onlookers describe the scene as a "feeding frenzy." Synonyms for the word *frenzy* are *hysteria*, *madness*, *insanity* . . . you get the picture: a chaotic free-for-all. In documentaries showing sharks feeding, the action is typically accompanied by music that reinforces this negative stereotype. To observe the feeding of the Caribbean Reef Sharks from the water evokes a very different description: *a peaceful dance.* The sharks move in and out of the feeding area, then use their senses, which have been honed over 450 million years of evolution, to locate the food, and then engage their prodigious swimming musculature to accelerate toward it. The only frenzy I see is among some observers who may become mad with delight, myself included.

Such is the nature of the double standard sharks face from the media and general public, who often vilify and demonize sharks instead of giving them the respect they are due.

Finning

It was not until 2010 that I developed a more complete picture of the extent of shark finning. During a stint teaching with the Semester at Sea program, while exploring Hong Kong, I found myself surrounded by a most extraordinary array of dried seafood shops. Displayed prominently in many of them were jars full of dried shark fins, selling at astonishingly high prices, as much as $225 USD per pound ($495 USD per kilogram).

Since the 1980s, the most economically valuable part of most sharks has been the fins, and they remain one of the most profitable. Shark fin soup is both a delicacy and status symbol in East and Southeast Asia, including China, Hong Kong, Taiwan, Singapore, Malaysia, Vietnam, and Thailand. Its first recorded use dates back to either the Song dynasty (approximately 1,000 years ago) or the Ming dynasty (600 years ago).

The presence of shark fins in a market or restaurant does not necessarily mean that they were obtained illegally, or that the sharks from which they came were not caught sustainably. It is entirely possible that shark fins in trade were legally caught and brought back to port attached to the body, which was used for meat and other products. The term *finning* is reserved for fins removed from the shark's body at sea, after which the lower-value carcass is dumped back in the water. It is worth noting that there are many more regulations aimed at preventing finning than overfishing. While finning is banned at many levels around the globe, there are still very few limits on shark catch.

In general, all shark fins except the upper lobe of the tail (caudal) fin are used for shark fin soup. After processing, which takes a week or more, what is left is the

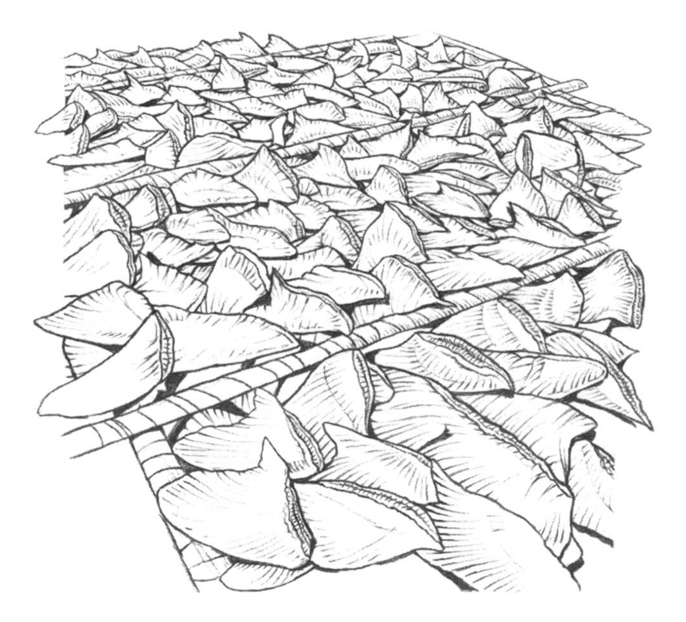

key ingredient for shark fin soup, the skeletal elements called ceratotrichia, known as "fin needles" in the industry. Ceratotrichia are made of a protein similar to the fibrous protein keratin found in human hair. After boiling, ceratotrichia resemble colorless cellophane noodles, which are then ready to be made into soup.

Fins from most sharks and some rays can be used for shark fin soup, with some of the most desirable fins coming from the hammerheads and requiem sharks, as well as wedgefish and guitarfish. Hong Kong remains the world's biggest trader of shark fins, with about half the global total, followed by Malaysia, China, and Singapore. Commercial fishers from Spain, Indonesia, India, Mexico, and the United States consistently report the largest shark catches.

While most international fishery bodies have finning bans, most rely on a complicated fin-to-body-weight ratio that is difficult to enforce. Conservationists are working to improve finning bans by adding requirements that sharks be landed with fins attached, as this is widely regarded as the best practice for finning ban enforcement. In 2022, the 186-nation Convention on International Trade in Endangered Species of Wild Fauna and Flora (CITES) added sixty species of requiem and hammerhead sharks, and thirty-seven species of guitarfish to Appendix II, thereby mandating that member governments document that trade is legal and sustainable. The CITES implementation process, however, is lacking. Moreover, as CITES measures do not directly affect fishing or finning restrictions, domestic use of sharks, or habitat conservation, it is a tool that should be complemented with others.

In the United States, trade in shark fins was banned as part of a $1.7 trillion USD omnibus spending bill, a move popular with conservationists in general but less so with many shark biologists, who think it will damage the already well-managed US shark fishery. This, in turn, may have the unintended consequence of encouraging increased illegal finning elsewhere to meet global demands. Whether this will happen is not yet clear.

Fisheries

Commercial fisheries in general increased rapidly in number after the end of World War II. Why then? Two reasons: a surplus of vessels that could be converted for fishing and the advent of sonar technology developed during the war that could be used to find fishes.

Although sharks and batoids have been used by humans for over five thousand years, it wasn't until the 1980s that shark fisheries experienced a similar post-WWII surge, due in part to depletion of some stocks of bony fishes. At about the same time, fresh shark meat became more acceptable as an alternative to Swordfish, tuna, and so on, in wealthier countries. A particularly pivotal date in the recent history of shark fisheries is 1972, when President Nixon initiated normalization between the United States and China, leading to the opening of a huge market in China selling shark fins for the delicacy shark fin soup.

Sharks and batoids (and other fishes as well) are caught using gear that falls into four general categories: active entrapment (seines, purse seines, trawls); hook and line (trolling, longlines); passive entrapment (trap nets, pound nets); and entanglement gear (fixed and drift gillnets, trammel nets).

Fisheries can be either *subsistence, artisanal, or commercial*. Subsistence fishers use traditional techniques and small vessels without advanced technology. Their catch is for the consumption of the fisher and his or her extended family. Subsistence fishing is thus vital for food security in some areas. At the other extreme, the most advanced, largest-scale fishery is commercial or industrial. Boats may be quite large, and are well-equipped, often with technologically sophisticated instruments for finding fish. Commercial fisheries are the most closely monitored, but illegal and undocumented fisheries occur with alarming regularity. Artisanal fisheries are intermediate between subsistence and commercial fisheries in practically every aspect.

Currently, sharks and batoids are caught for their fins, meat, liver, cartilage, skin, teeth, and jaws, as well as other body parts. As of 2022, the market for shark meat is still growing.

The largest importers of shark and batoid meat include the Republic of Korea, Spain, Italy, Brazil, and Uruguay. Korea's imports are predominantly skates, which are used in a traditional fermented dish known as *hongeo-hoe*. Although trade in shark fins peaked in the early 2000s, the overall value is close to that of shark meat.

As a group, shark and batoid fisheries are less than 1 percent by weight of total marine capture fisheries. Shark fisheries peaked in 2000, and since then have declined. For 2018, the global shark and batoid catch was only about 0.7 percent of total global fish catch, a drop attributed to declining stock of sharks and batoids as well as implementation of effective management that included the establishment of marine protected areas or sanctuaries, quotas, and so on.

In the United States, catch of sharks and rays peaked in 1992 and has declined since then. Catch reductions in the US are due to a combination of decreased abundance from overfishing, implementation and enforcement of fishery management plans, and falling market demand. Some populations, like Gulf of Mexico Blacktip Sharks and New England Winter Skates have recovered under science-based catch limits while others, such as Atlantic Dusky Sharks and Thorny Skates remain seriously depleted despite decades of protection.

The main problem with shark and ray fisheries has been insufficient protections, such as catch limits, despite life history characteristics, such as slow growth and low fecundity, that make the fished species exceptionally vulnerable to overfishing. Sharks and rays have been low priorities for fishery managers due usually to a combination of relatively low economic value and lack of public support. At the same time, necessary protections are also hindered if the associated cost to fishermen is considered too high.

A major problem has been the sale of fishing rights by island nations and smaller poor countries to big fishing nations like Korea, Japan, India, Spain, and so on, which typically overfish all stocks. Another major problem is burgeoning deep-sea fisheries for sharks that have delayed life history characteristics, such as slow maturation and limited fecundity and thus are more vulnerable to overfishing.

Despite the seemingly overwhelming negative impacts of shark and batoid fisheries due to a toxic blend of overfishing and slow life history characteristics leading to low potential to recover quickly, sustainable shark fisheries are possible and do indeed exist, such as Dusky Smoothhounds in the US (despite the weakest finning controls in the country) and Gummy Sharks in Australia.

Gnathostomes: The Rise of Jaws

We take our jaws for granted, but the earliest known vertebrates, *ostracoderms* from 460 million years ago, were jawless, and they prospered for perhaps 100 million years. How did they eat? Ostracoderms apparently sucked up small, soft-bodied invertebrate prey from the bottom. Ostracoderms gave rise to the *agnathan* (literally meaning "without jaws") hagfish and lampreys, a group that has persisted nearly 500 million years, but with limited success: only about 122 species remain today, compared to the approximate 38,000 bony fishes and nearly 1,250 elasmobranchs. Why so few jawless fishes today? The rise of jaws was monumental, so much so that it constrained the diversification of the jawless fishes. Jawless fishes were simply outcompeted by these showy new kids on the block with their swanky maws.

Jaws arose much like other specializations, through natural selection. In the first fishes to evolve jaws, the acanthodians and placoderms, the first set of gill arches transmogrified into the upper and lower jaws over evolutionary time, and these were structurally supported by modifications of the second set of gill arches. These

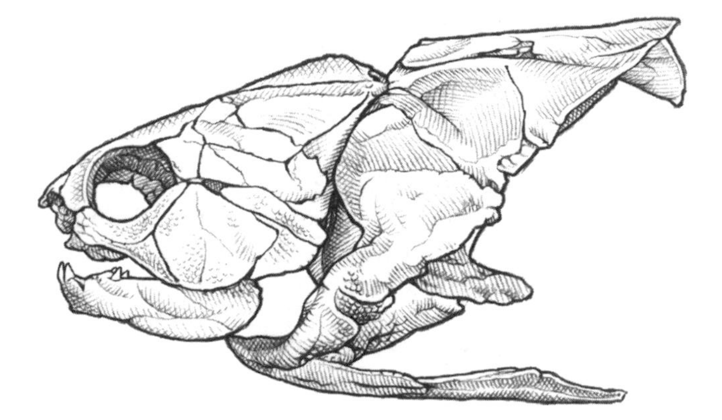

fishes, thus, were the first gnathostomes (*gnath* meaning "jaws"; *stome* meaning "opening"), and all jawed vertebrates are sometimes classified as gnathostomes.

The advantages of jaws may not be obvious to you, but being able to grasp prey, and manipulate and process it before swallowing, enabled the diversification of the type and size of suitable prey. It is not hyperbole to aver that the rise of jaws arguably is the key event in the vertebrate success story, and that it left the jawless fishes gasping for air, so to speak.

Golden Age of Sharks

From the perspective of the widespread interest in all things shark among the public, the present era could justifiably be labeled the Golden Age of Sharks. This honorific, however, has already been taken; it belongs to the Carboniferous Period (300–360 million years ago), when early sharks and their chondrichthyan relatives experienced their first major radiation (large increase in diversity). During this period, there were more sharks and rays than any other fishes, and they were the dominant predators in oceans, rivers, and lakes. But numbers tell an incomplete story; the diversity of these sharks was astonishing and included numerous bizarre species with features such as unicorn-like heads and scissor jaws.

Alas, all good things must come to an end. In this case, about 250 million years ago, the Great Permian Extinction occurred, during which 90–95 percent of all marine species, sharks included, were wiped out. Some shark lineages survived, especially those living in deeper waters.

Great Hammerhead

Otherworldly. Alien. Unreal. These and other adjectives are often used to describe this shark, one of nine species in its family, all of which feature the eponymous, oddly shaped head. The Great Hammerhead is distinguished from some of its confamilials (family members) by the absence of scalloping or a smooth curve on the leading edge of the head, and the presence of an enormous first dorsal fin, large pelvic fins, and a huge upper caudal fin.

Great Hammerheads, which can reach a length of 20 ft (6 m), are distributed throughout tropical and warm, temperate seas in both inshore and pelagic environments, from the surface to 1,000 ft (300 m). They eat a variety of bony fishes, as well as other sharks and marine invertebrates. They also prey on rays. Their weird head, called a cephalofoil, may play a role in both maneuverability and stability. Hammerheads have also been observed using their heads to pin rays to the bottom before eating them.

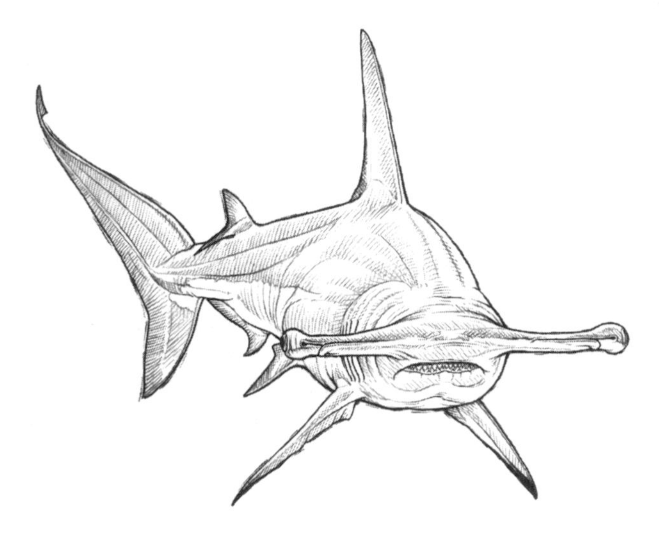

The Great Hammerhead is classified by the IUCN as "critically endangered." Capture mortality is very high in this species; a recent study showed that 50 percent were dead after three hours on a hook.

And this just had to happen: the artist Matt Sanders sculpted a hammerhead shark using exclusively—you guessed it—the heads of hammers, and displayed his creation at the Aquarium of the Pacific in California.

Great White Shark

What can be added here about perhaps the most iconic of all the sharks, the White Shark? For starters, the "official" common name is, in fact, White Shark and not *Great* White Shark.

My first interaction with the species was in the early 1980s while working on my PhD at Scripps Institution of Oceanography in La Jolla, California. In exchange for helping San Diego's Sea World understand the physiology of the species in captivity—specifically, why the species did so poorly—Sea World helped with my study of heart function in sharks.

The answer to Sea World's question embodies the essence of what makes the White Shark, as well as its four mackerel shark relatives, the astonishing predators that elevate them in our fascination.

Central to the White Shark's prowess as a powerful predator is a metabolic engine that enables the species to stay warmer than its environment, which translates into increased muscle power and provides fuel to feed their powerful, energy-demanding swimming muscles. The White Shark's streamlined physique and sickle-shaped

caudal fin are the other ingredients that make the White Shark the preeminent predator of the shark world, and one of the top oceanic predators of any group.

To complete the story about Sea World: We determined that captive White Sharks typically succumb to a suite of physiological problems due to capture stress. These included toxic build-up of chemical by-products of metabolism, overheating, salt and mineral imbalances, water imbalance, and hypoxia (low internal oxygen levels).

Among public aquariums, only the Monterey Bay Aquarium in California has been successful in maintaining the species in captivity, albeit for limited durations. Their record, in 2004, was 198 days, after which the shark was released back into the wild.

An interesting story about White Sharks is the discovery of an area about the size of New Zealand midway between Hawaii and the Baja Peninsula of Mexico, where White Sharks are known to dawdle. This zone has been called the "White Shark Café." The White Sharks there remain near the surface during daylight and dive as deep as 1,500 ft (450 m) at night. What are the sharks doing there? Though the area was named café because the sharks lingered there, it turns out the appellation was accurate: the sharks are likely taking advantage of unexpectedly dense concentrations of prey and dining there (though they are not sipping oat milk lattes).

White Sharks can reach 20 ft (6 m) and 4,200 lb (1,900 kg). They have a jaw full of triangular serrated teeth for cutting, long gill slits, very black eyes, vivid color changes on the sides, and black tips under the pectoral fins. White Sharks are found globally in temperate and tropical coastal and oceanic waters. Contrary to the public's perception, the overall population is not declining, although some regional populations may be. Stay tuned for updates as human impacts, particularly climate change, affect them noticeably.

Happy Eddie Shyshark

Most common names are merely descriptive, like Tiger Shark, Blue Shark, or Pyjama Shark. There is one species, however, whose common name is partially descriptive but wholly, categorically whimsical: the Happy Eddie Shyshark. Also called the Puffadder Shark, a reference to its coloration, which is similar to that of the Puff Adder snake, the Happy Eddie Shyshark inhabits coastal waters of southern Africa. It belongs to the catshark family.

How did this species acquire its cheerful moniker? Here's a hint: look at its scientific name. Happy Eddie is a diminutive of *Haploblepharus edwardsii*. Why is it a *shy* shark? Because, when it is threatened by predators like Cape Fur Seals, it has a habit of rolling into a ball with

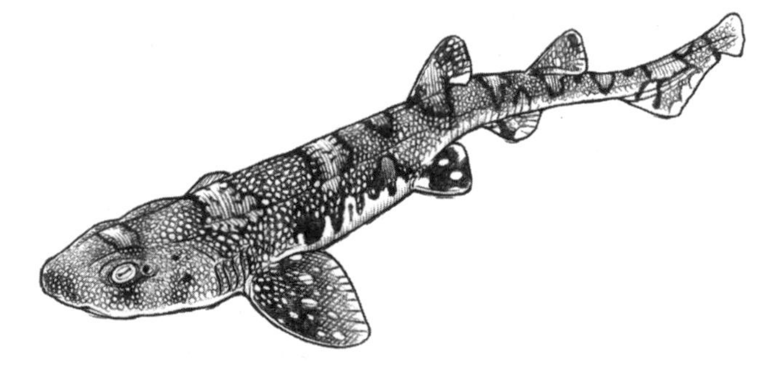

its tail covering its eyes. For that reason, the species is also called a Donut Shark. Maybe in doing so, the shark encourages the often playful Cape Fur Seals to engage in a game of "toss the shark" rather than eating it.

Other interesting common shark names abound, including the Dumb Gulper Shark, Ninja Lanternshark, Pigeye Shark, Nervous Shark, Pinocchio Catshark, Lollipop Catshark, Phallic Catshark, and lastly, in an obvious affront to the so-called Great White Shark, the Magnificent Catshark.

Heart Anatomy and Physiology

In the early 1980s, as a doctoral student in southern California, I was asked to provide some questions and answers to a television game show, *Hollywood Squares*, as I recall. The producers provided me with the first question: *who has a larger heart, a human or a white shark?* The answer, of course, depends on the relative sizes of both organisms. An adult human has a heart about twice as large as that of a similar-sized White Shark, but the heart of a full-grown White Shark dwarfs that of the adult human. I once dissected the heart of a 2,100 lb (950 kg) White Shark that had died in the gill net of a commercial fisher; it was huge, nearly the size of a volleyball, and it weighed almost 6 lb (2.7 kg).

The relative mass (that is, heart weight divided by body weight) of shark hearts varies among species, and it correlates strongly with activity level, as it does generally among vertebrates. Thus, the high-performance mackerel sharks, such as the Shortfin Mako, have the largest hearts and the relatively sedentary species, like the Horn Shark, have the smallest.

Notwithstanding the enormity of the heart from the White Shark I dissected, the hearts of the sharks I have dissected never cease to stun me with their diminutive proportions. How could an organ smaller than a ping pong ball circulate the blood throughout the body of a 3.3 ft (1 m) Blacktip Shark? The answer is that it is indeed a powerhouse of a miniature pump, but it receives an assist from the movement of the posterior trunk of the shark, which helps propel the blood from the rear of the body back to the heart, which is located in the ventral part of the shark just posterior to the gills. Aquarists understand this, and when transporting some sharks, will move the tail end of the shark as if it is swimming, which advances the blood to the heart and prevents it from pooling in the lower and rear body and rupturing small blood vessels there.

Here's another interesting and, to a shark biologist, maddening fact: almost all introductory and even some advanced biology textbooks authoritatively but mistakenly state that the heart of sharks has only two chambers. While it is true that the heart of sharks has only a single atrium and ventricle, and mammals (including you) have two of each and thus have a four-chambered heart, sharks have two additional contractile chambers that any anatomist would conclude are authentic heart chambers. Sharks thus have true, four-chambered hearts. Mammalian chauvinism and arrogance continue to rear their ugly heads. Like humans, sharks have coronary arteries that nourish the heart muscle.

Finally, one of my favorite pastimes in port cities is to visit fish markets, which are windows into the culture, economy, diet, and so on, of that area's residents.

In fish markets in Asia, I was surprised to see displays of dozens of shark hearts. Like many organ meats, it is cut into small pieces (after draining the blood), which may be dipped into a sauce and consumed raw.

Today's lesson: sharks are not heartless after all!

Jaws Phenomenon

Shark biologists have mixed feelings about Peter Benchley's book *Jaws* and the 1975 blockbuster movie of the same name. On the one hand, *Jaws* jump-started an interest in sharks that has exploded since then. On the other, this interest has manifested itself fundamentally in an increased fear of sharks and a media empire that serves predominantly to inflame these fears.

Although the book was a bestseller, Steven Spielberg's 1975 movie, often called the *first* summer blockbuster movie, is credited with catalyzing the ensuing frenzy.

The depiction of the central character—by which I mean the White Shark (a mechanical beast nicknamed Bruce, after Spielberg's lawyer) in the original *Jaws* as well as in sequels—as a vengeful eating machine with a particular taste for humans is, of course, pure fiction. But the movie struck a chord that writers from Bram Stoker to Mary Shelley to Stephen King and beyond have successfully exploited—our innate fear of so-called monsters.

Although author Benchley denied any connection, the plot of *Jaws* is linked to a series of shark attacks in 1916 along the Jersey Shore. During a twelve-day period that summer, four people were killed and one was injured. Interestingly, at the time, many ichthyologists dismissed that sharks were responsible. Experts instead conjectured, almost hilariously if not for the seriousness of the injuries, that the list of perpetrators might include German U-boats, sea turtles, and even Orcas. What is even more interesting is that, over a hundred years later, there is still uncertainty about what kind of shark, a White Shark or a Bull Shark, was responsible for the attacks (but we are fairly certain that it was not an Orca, sea turtle, or U-boat).

One additional comment. You might recall this iconic scene from the movie *Jaws*: With the sounds of seagulls in the background and a gentle wind blowing under blue skies, a group of carefree adults and kids are frolicking close to shore when one of them shrieks in terror and the lifeguard blows his whistle, leading everyone to chaotically run from the water amid a chorus of screams. Then silence, as the stunned people on the beach stare seaward and a deflated raft drifts ashore, with an arc of fabric removed by the White Shark where a child had been. Filmmaker Lesley Rochat, an award-winning shark conservationist and good friend, has produced a variation of this scene (available on YouTube) to highlight our misconception about the probability of shark attacks. In her version, the focus is not the remnants of the raft but instead a toaster bobbing at the surface, followed by the superimposed message: "Last year 791 people were killed by defective toasters, 9 by sharks. Rethink the shark."

Postscript: In December 2022, filmmaker Steven Spielberg apologized for the negative impact his movies had on shark populations. Author Peter Benchley, who died in 2006, had years earlier similarly regretted his sensationalized depiction of sharks, and began working for shark conservation.

Jumping Sharks

It may be a surprise to you that numerous species of bony fishes, primarily those living in the tropics, are capable of breathing air or living amphibiously (mud-skippers, for example). No sharks can do either of these things; all are obligate aquatic water-breathers. However, several species of sharks are capable of becoming aerial,

if only for a heartbeat or two. In addition to the prodigious, death-defying leaping ability of White Sharks in pursuit of marine mammal prey, numerous other species of sharks are known to make equally spectacular jumps, including Blacktip Sharks, Spinner Sharks, Basking Sharks, Shortfin Makos, threshers, and doubtless others.

Anyone who has witnessed the unexpected appearance of a large gray beast of a shark doing aerial acrobatics at eye level will never forget the jaw-dropping, indescribable spectacle that leaves witnesses with a *what-just-happened?* expression.

Most surprising regarding leaping sharks is the 2018 research report of the least likely candidate to muster the high energy to reach the required escape velocity in order to break the surly bonds of water: the languid, humongous, filter-feeding Basking Shark.

Explanations for jumping in sharks include feeding, intraspecific communication, dislodging remoras (also known as sharksuckers), removing ectoparasites, and reacting to water temperature changes. To observe the phenomenon leads to an additional inexorable, anthropomorphic explanation: the sharks are merely expressing their exuberance!

Life History Characteristics/Vulnerabilities

As a group, sharks have slow, or conservative, life history characteristics, and some resemble those of an African Elephant. Most shark species exhibit slow-growth, long life spans, late sexual maturity, long gestation periods, low fecundity, and reliance on specific mating and nursery areas. These life history characteristics have suited sharks well throughout their evolutionary history of approximately 450 million years. However, when added to the relatively small population sizes of many shark species, you have the perfect storm of vulnerability to human insults like overfishing, pollution, habitat alteration, and so on.

Organisms employ one of two main life history strategies, or exist along the continuum between these extremes, in their evolutionary journey. On the one hand are organisms like fruit flies, which grow quickly, mature early, and lay lots of small, quickly hatching eggs. At the other end of the spectrum are those that include large terrestrial mammals and sharks, which may take a decade or longer to reach sexual maturity and have very low fecundity (as low as a few offspring at every reproductive event, which may occur every other year, or even every three years).

Let's examine a few of the life history characteristics of sharks.

— Sharks are slow-growing. It takes female Sandbar Sharks about fifteen years to grow from 16–20 in (45–50 cm) at birth to sexual maturity at approximately 54 in (135 cm), an average growth rate of just 2.4 in (6 cm) per year.
— Sharks are generally long-lived and reach sexual maturity late in life. Sandbar Sharks live 30–35 years, and Greenland Sharks may live four hundred years or more! Spiny Dogfish take as long as twenty years to reach sexual maturity.
— Gestation (period of pregnancy) is also slow going: about a year for Sandbar Sharks, as long as two years for a Spiny Dogfish, and three and a half years for Frilled Sharks!

— Fecundity (number of offspring) is typically low. Sandbar Sharks produce about eight pups every two years.

Some sharks break the above rules. Tiger Sharks may produce sixty or more pups. Skates and catsharks can produce dozens of offspring through egg-laying, whereas, because of their diminutive sizes, they could produce only a few similarly sized offspring if they utilized live birth. Egg-laying in sharks and rays is at least in part an adaptation for increasing fecundity in small species.

Why do these life history characteristics matter? For one thing, they make it difficult for shark populations to quickly recover from perturbations like overfishing, whereas a Yellowfin Tuna, more fruit fly than shark in terms of life history, matures at age three and lays millions of eggs, and thus can theoretically recover quickly from overfishing if effectively managed (say, in only a few years).

It is critical for scientists to learn about life history characteristics for all sharks and for managers to restrict fishing accordingly to conserve populations. When in doubt, a precautionary approach is warranted. Until that happens, many species will remain at risk.

Life's a Drag: How Sharks Swim

In the movie *Annie Hall*, the character Alvy Singer says, "A relationship, I think, is like a shark, you know? It has to constantly move forward or it dies. And I think what we got on our hands is a dead shark." Notwithstanding the fact that this rule does not apply to benthic sharks, Alvy was basically saying *sharks gotta swim*.

Moving through water is not easy. It requires that the water be pushed aside, which is a problem since water is so dense and, compared with air, for example, is resistant to flowing. Sharks, however, have evolved pathways to successful swimming in spite of these drawbacks.

First, there is the idea of balance. Any object moving through a fluid may rotate in three ways. *Yaw* involves side-to-side movement, typically of the head. *Pitch* is the up-and-down movement of the body, like a boat heading through a series of waves hitting it head on. *Roll* is the rotation of the body, which can make a shark go belly-up. These motions are managed mostly by the fins. Perhaps surprisingly, however, the precise role that fins play in achieving balanced movement is not as well understood as one might think.

Let's talk about the caudal fin, which provides thrust. Due to the asymmetry of the caudal fin, the tail is thrust diagonally upward. If sharks had wimpy pectoral fins, then, in the same way that as one end of a seesaw goes up when the other end goes down, every stroke of the caudal fin would lift the tail and push the head down, and sharks would turn summersaults as they moved forward—a case of the tail wagging the dogfish.

But sharks have stiffened, larger pectoral fins, and in many cases horizontally expanded heads, both of which may act in combination with their anterior body not only to resist any downward forces but also to add lift so that sharks swim in a controlled, balanced, straightforward way.

The name of the game in swimming is overcoming the forces that work against forward movement, collectively called "drag." Although this explanation is oversimplified, drag is reduced in sharks through streamlining and the structure of the dermal denticles.

Most fish, including sharks, swim by undulating their body in combination with oscillating their caudal fin. Among sharks, there is a high diversity of swimming styles and speeds.

There are three major modes of swimming in sharks (named for the bony fishes that best exemplify the mode). In *anguilliform* (eellike) swimming, found in slim, elongate sharks such as the Frilled Shark and catsharks, an undulatory wave moves down along the entire body, as with the movement of eels. The swimming mode of most sharks is called *carangiform* or *subcarangiform* (named after swimming in the jack family). Here, the undulations occur mostly in the posterior half of the body. The final category is *thunniform* (tuna-like), on the opposite extreme from anguilliform. Like a windup fish toy, only the caudal fin and peduncle oscillate, and there is little head yaw. Among sharks, thunniform swimming is restricted to members of the family Lamnidae.

One final, cool swimming adaptation is the Sand Tiger's hovering. In aquariums, the Sand Tiger inches forward as if swimming through molasses, nearly hovering. How, you might marvel, can such a hefty beast maintain its position in the water column without moving more speedily? An internal balloon, that's how. Sand Tigers will gulp air at the surface, and the lining of the stomach traps the air bubbles. In so doing, Sand Tigers create what other sharks cannot: they make their own ersatz bony-fish swim bladder. Using a floatation device saves energy, since it is much more energetically costly to move under your own power than to float, and energy efficiency is critical to evolutionary success.

Living Fossil

Sharks are commonly called *living fossils*, organisms that resemble their ancestors. Horseshoe crabs, crocodiles, and even ferns are considered living fossils. To the extent that sharks settled on a body shape and predatory lifestyle that was successful early in their approximately 450-million-year evolutionary history, and that some aspects of these have persisted to the present, then calling them living fossils is justifiable.

Other characteristics of sharks, however, have changed significantly over evolutionary time, challenging that perception. For example, the jaws of modern sharks have become more protrusible and have a larger gape than those of their ancestors, enabling them to grasp, shear, and manipulate their prey in ways that would be impossible with a fixed jaw like their early ancestors (or like your jaws).

Additionally, the fins of modern sharks, though still stiff, have become slightly more movable, and they thus differ from the more rigid fins of early sharks. This change makes modern sharks more maneuverable than many of their ancestors.

Finally, bony fishes like mackerel and others trace their origins back to the same geological period as sharks, so if sharks are living fossils, so are bony fishes! Thus, modern sharks are not primitive creatures but are advanced and well-adapted beasts. Plus, modern sharks do not generally resemble their early ancestors.

egalodon/Megatooth

One of the coolest fossil finds is a large tooth from the extinct shark commonly known as

Megalodon (*Otodus megalodon*), the largest predatory fish ever to exist, reaching up to 60 ft (18 m) in length. For comparison, White Sharks have been recorded as large as 20 ft (6 m) and 4,200 lb (1,900 kg). Megalodon, which lived from 23.5 to about 2.5 million years ago, is mistakenly considered a direct ancestor of the White Shark. As satisfying as this story would be—that the biggest predatory fish of all time gave rise to the most feared shark alive today—it is apocryphal. Megalodon is an evolutionary dead end and did not give rise to any marine animal living today. Based on dates of its fossil

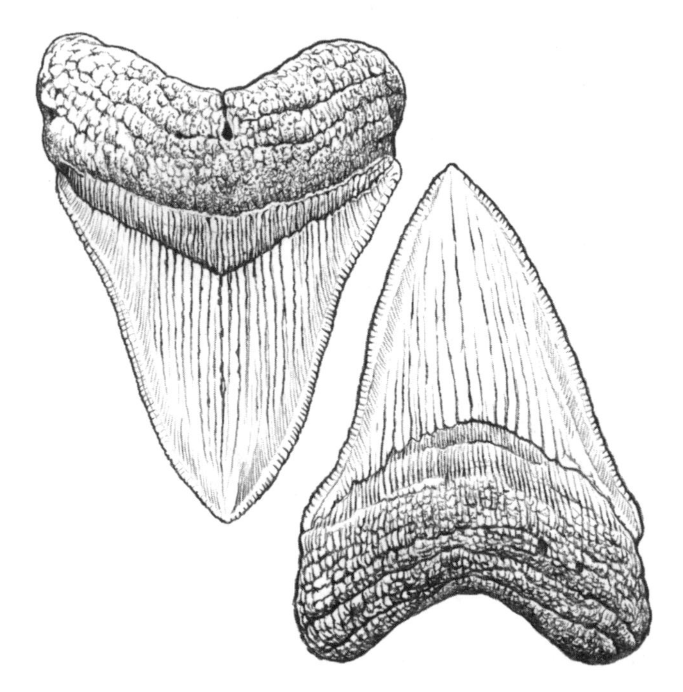

teeth and vertebrae, the White Shark genus *Carcharodon* dates back 60–65 million years. Megalodon is thought to have arisen from a relative of the mako sharks, and White Sharks may have helped drive Megalodon's extinction through competition.

But that tooth is still *way* cool.

Migrations (Movement Ecology)

Many sharks migrate, and some are *highly* migratory. Shark migrations may occur on short-term timescales, such as tidal or day-night (diel) cycles, intermediate scales like lunar cycles, or longer, seasonal cycles. The migratory distances vary as well, from a few to hundreds or even thousands of miles.

The major exception are those sharks that do not migrate over long distances: deep-sea sharks. Why would they? Their environment is characterized by biological, physical, and chemical variables that basically do not vary with tides, time of day, or even seasons. Since migrations often are associated with environmental conditions, resource availability, finding refuge, mating, or giving birth, why waste energy and time migrating to areas where conditions are unchanged from those of your current habitat?

Some deep-sea sharks do migrate over shorter time and distance scales, as part of the largest migration on the planet, the diurnal vertical migration from the deep sea to the epipelagic zone near the surface at night. The migration involves small fishes, squid, larger fishes, and mesopelagic sharks such as Cookiecutter Sharks, Goblin Sharks, Frilled Sharks, Gulper Sharks, Cuban Dogfish, and Bluntnose Sixgill Sharks. This group ascends from the depths at sunset to feed on the accumulation of zooplankton and phytoplankton, then migrates back at sunrise. To sharks and other predatory animals of the deep that do not migrate, the migrators represent the Uber Eats of the depths, delivering the newly acquired food in their stomachs (as well as themselves) to the stomachs of the lazy and patient—but clever—non-migrators.

Sharks migrate for one or more of four reasons: physiological tolerances to environmental factors such as temperature or salinity, prey availability, reproduction, and refuge.

Consider Sandbar Sharks, which can be used as examples of all four types. In the late fall in Chesapeake Bay, when water begins to chill to levels that juveniles cannot tolerate (approximately 41° F, or 5° C), they migrate to warmer waters. A different population of juvenile Sandbar Sharks migrates to Winyah Bay in South Carolina in the spring, where they use lower-salinity parts of the system for both food and refuge from larger predators. Finally, also along the East Coast of the United States, adult Sandbar Sharks migrate to mate.

Noteworthy additional examples of navigation include migrations by Spiny Dogfish of up to 4,400 mi (7,000 km). Not bad for a small shark! Blue Sharks migrate from the southwestern to the southeastern Atlantic Ocean. Along the US East Coast, large schools of Blacktip Sharks appear along the coastline of south Florida around the middle of January and stay there for about two months before moving north. Generally, benthic sharks and smaller sharks do not undertake long migrations.

Migratory cues for sharks include seasonal changes such as day length (photoperiod) and temperature, and the relative importance of these can vary by species. For example, Sandbar Sharks in the Chesapeake Bay rely more on day length changes whereas Blacktip Sharks along Florida's west coast respond more to temperature changes.

Migrating sharks often return to the same locations and remain in the general area. In other words, they exhibit *site fidelity*. A large variety that includes Oceanic Whitetip Sharks, Horn Sharks, Sandbar Sharks, and numerous others, do this. Many, including Lemon Sharks, are also *philopatric*, that is, they return to their exact or regional birthplace either to give birth themselves or to mate.

A 2021 study found that Bonnetheads used the Earth's magnetic field as a map for homeward orientation in their annual migrations, the first demonstration of how sharks navigated during longer migrations.

Most Magnificent Fish in the Sea

What attributes would be required for a fish to be given the honorific of Most Magnificent? Here are a few: striking coloration, a body shape that screams *paragon of efficient design*, unmatched predatory prowess, and prodigious metabolic machinery that enables it to outrun all of its prey. Lesser fish would fear it, if you will excuse my anthropomorphism. And it would have a suite of cool adaptations that other fish could only dream of.

The short list of candidates for the title would be limited to the mackerel sharks—that is, the five members of the family Lamnidae. These are the White Shark,

Shortfin Mako, Longfin Mako, Salmon Shark, and Porbeagle. A convincing case could be made for any of these, but one species stands above the rest, and it probably is not the one you envision: it's the Shortfin Mako.

The key adaptations of the Shortfin Mako that would justify its selection as the "Most Magnificent Fish in the Sea" include a streamlined, almost cylindrical body; striking countershading, with blue on top and white on the underside, which affords an additional measure of stealth when foraging for prey; a caudal fin designed for maximum forward thrust, with a flattened keel at its base, which together facilitate high-speed swimming;

a metabolic engine and anatomical specializations that keep the body warmer than the water in which it resides, enabling more powerful swimming muscles; a high-performance cardiovascular system to make sure the swimming muscles get the energy and oxygen they need and remove wastes; ram ventilation (swimming with its mouth open) and big gills with large slits to feed its oxygen-thirsty muscles; spindle-shaped teeth to grab and hold slippery fish; and a well-developed sensory system and a large brain for collecting and processing sensory information.

The defense rests.

Music

If you ignore the theme from the movie *Jaws* and you are at least as old as a baby boomer, only a single song qualifies as the most recognizable song with prominent mentions of sharks in its lyrics: "Mack the Knife." The relevant lyrics are:

> Oh, the shark has pretty teeth, dear,
> And he shows them, pearly white.
> When the shark bites with his teeth, dear,
> Scarlet billows start to spread.

Never mind that this song, most memorably sung by Louis Armstrong, Ella Fitzgerald, and Bobby Darin and composed in 1928 for the drama *The Threepenny Opera*, was not about a real shark at all but instead described a hitman. Sigh, yet another metaphor unfair to sharks.

In fact, most songs that include sharks in the name or lyrics use sharks as metaphors for other themes.

Additional adult songs about sharks other than the *Jaws* theme and "Mack the Knife" are few and far between. There is David Roth's "Hammerhead Shark" and Jimmy Buffet's "Fins." In the reggae world, there is "Shark Attack" by the Wailing Souls. And in the jazz genre, have a listen to "Baby Jazz Shark" by the European Jazz Trio.

Speaking of shark songs, a cool study by Andrew Nosal and colleagues from Scripps Institution of Oceanography at the University of California, San Diego, and Harvard, entitled "The Effect of Background Music in Shark Documentaries on Viewers' Perceptions of Sharks," concluded that attitudes toward sharks are influenced or reinforced by how ominous the background music is. Don't you just love science?

The most famous shark-inspired singer? Why Sharkira, of course. And the absence of the song "Baby Shark" here is intentional.

Nurseries

Young sharks, even those born at a large size, require protection from predators as well as adequate prey. This is where shark nursery grounds play a role. Places where sharks give birth may be *primary* nurseries, whereas discrete areas where young sharks spend a disproportionate amount of time may be *secondary*.

Use of nurseries by sharks likely evolved early in their evolutionary history, since the advantages of abundant food and protection from predators, as well as reducing competition between age classes of the same species, would seem to be so advantageous that to forgo them would be a liability.

My research team recently discovered a previously unknown secondary nursery for Sandbar Sharks in Winyah Bay, in northeast South Carolina. There, specifically in the middle regions of the bay, where the water is brackish, the eight or more species of adult sharks less than a mile away are physiologically incapable of tolerating this brackish water for long. But these juvenile Sandbar Sharks, dominated by the age class of 4–6 years, have evolved a physiological mechanism that allows them to inhabit this section of the bay for extended periods.

The bad news is that changing precipitation patterns caused by climate change, particularly droughts and deluges that cause freshets (influxes of fresh water over short periods), are changing the community structure of Winyah Bay, which may lead to it being less suitable as a secondary nursery, which may in turn hinder the recovery of this species.

Oceanic Sharks

If you think the deep sea has conditions that make living there difficult, you haven't visited the epipelagic zone of the open ocean, the part shallower than 660 ft (200 m). Fewer than 5 percent of sharks, approximately twenty-five species, are oceanic, and a large number of these also spend time below the epipelagic zone, in the mesopelagic zone.

True oceanic sharks include many familiar larger sharks, such as the Blue Shark, Oceanic Whitetip Shark, Shortfin Mako, and Common Thresher, plus a few with which you may be unfamiliar, like the Longfin Mako and Bigeye Thresher. Additional oceanic species include Silky Sharks, Salmon Sharks, and Porbeagles.

Why such low diversity? The open-ocean ecosystem as a whole is not as productive as coastal ecosystems, which translates into less biomass (think "prey") and a lower biodiversity of large predators, including sharks. The lack of barriers in the open ocean means little habitat diversity and fewer ecological niches compared to coastal habitats like estuaries or coral reefs. It also encourages the shark inhabitants to be wide-ranging, and many are thus also distributed circumglobally.

Many sharks of the open ocean also exist in a single population for an entire ocean basin. For example, there is only one population of Blue Sharks across the entire North Atlantic Ocean. In contrast, Blacktip Sharks, a coastal species, have at least three genetically distinct, noninteracting populations along the Atlantic and Gulf coasts in the US. From a conservation perspective, the wide distribution is disadvantageous, in that it exposes species to fisheries from many countries.

Finally, since food resources are scarce in the open ocean, sharks there are food generalists. When they encounter something of an appropriate size or shape, they must ask themselves when they might have another opportunity to eat. This explains the diverse array of items found in the digestive tracts of oceanic sharks, such as suits of armor, tin cans, car tires, and so on.

Parental Care in Sharks

There is none. Isn't it virtuous enough that sharks do not eat their offspring?

Parthenogenesis (Virgin Birth)

Vertebrate reproduction typically requires a female whose ova are fertilized by the sperm of a male. An alternative reproductive phenomenon that produces young that are both viable and fertile (capable of surviving and, when mature, reproducing) is *parthenogenesis*, or virgin birth. In many groups (e.g., in some plants, insects, and bony fishes), parthenogenesis occurs as a natural form of asexual reproduction, but it was unknown in sharks until reported for a captive Bonnethead. Parthenogenesis has now been described in multiple species of captive sharks as well as one species of ray. There is also one report of parthenogenesis occurring in the wild and leading to viable offspring in the Smalltooth Sawfish.

The adaptive value of parthenogenesis is thought to be this: When males are scarce or are separated from females, a new generation of young can still be birthed.

A major disadvantage is that parthenogenesis reduces genetic diversity, the raw material for evolution.

Planktivorous Sharks

All sharks are predators, but not all sharks are top predators. Among sharks at the other end of the food chain, three species have evolved to assert their predatory prowess as planktivores (that is, plankton-eaters): the Whale Shark, Basking Shark, and Megamouth, all of which are the only members of their separate taxonomic families.

All three planktivorous sharks are slow-moving, thus requiring less food than top predators, and they eat lower on the food web, where there is greater prey biomass and thus more energy.

The largest *animal* ever to inhabit the planet is the Blue Whale, and it is planktivorous.

The Whale Shark is the planet's largest extant *fish*, up to 59 ft (18 m), and it is wide-ranging in tropical and warm, temperate waters. It feeds by ram filtration, opening its mouth and filtering plankton and small fishes as it slowly swims. It also employs suction by expanding its oral cavity and inhaling large volumes of water, which it expels through its gills after closing its mouth. There are twenty or more areas off the coasts of Australia, Belize, the Maldives, Mexico, the Philippines, Qatar, India, China, and others that are considered Whale Shark hotspots because of their large aggregations. Whale Sharks are killed for meat, liver oil, cartilage, and especially their large fins, which are exported. Unbelievably, Whale Sharks are maintained successfully in captivity, even after being shipped as far as 8,000 mi (12,900 km)!

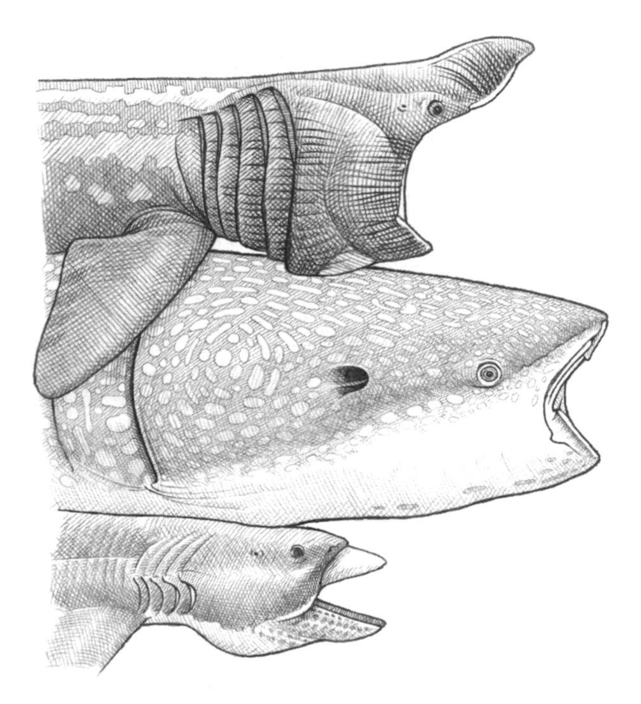

The second largest shark (and fish as well) is the Basking Shark, at 33 ft (10 m). It is found worldwide in cool water. They have huge gill slits almost encircling their head, and they filter plankton with their gill rakers, which are deciduous, falling out and being replaced annually. Basking Sharks are one of the poster children for overfished sharks. Until the mid-twentieth century when their populations crashed, the species supported a large fishery in the North Atlantic. The IUCN now lists them as "vulnerable," since they have been on a recovery trajectory. Fun facts: Basking Sharks have small brains relative to their body weight, probably a function of the weak swimming and diminished sensory requirements of a planktonic lifestyle. They recently have been observed leaping from the water along the coast of Ireland, the shark equivalent of a tortoise speed-walking.

The Megamouth Shark was unknown to science until 1976, when one was caught enwrapped in a sea anchor. The next specimens were not caught until 1984 and 1987. As of 2022, there have been over 270 documented captures or sightings, most from Taiwan. The Megamouth Shark feeds by swallowing water and then forcing it forward through the gill rakers, which strain out the plankton. The species is probably found worldwide in the tropics, with fewer in the Atlantic.

Plastics in Sharks

We live in the Age of Plastics. As much as 1.8 and 10 percent of the annual global plastic production, which is over 660 billion lb (300 billion kg), enters the ocean, as either macroplastics or microplastics. The former category includes beverage bottles and packaging as well as abandoned commercial fishing gear. Microplastics are less than 0.2–0.4 in (5–10 mm) in size and may be produced as degradation products of larger plastics, or may be purposefully manufactured as microplastics—for example, microbeads in cosmetic and personal care products.

Once in the marine environment, plastics have been shown to cause a suite of problems at the organismal and ecosystem level, although much remains to be understood about these impacts, especially in sharks. Broadly, problems are associated with either entanglement or ingestion. I have caught (and thankfully was successful in releasing), Sandbar and Blacktip Sharks entangled in plastic strapping material.

Microplastic toxicity may be due either to adsorption of nonpolar chemical pollutants to the surface of the microplastics (e.g., PCBs, DDE, and DDT), or harmful chemicals in the plastics (e.g., flame retardants and plasticizers). Surprisingly, studies on the effects of toxic microplastics on sharks and batoids are lacking.

Given the pervasiveness of plastics in the marine environment, the continued growth in their production and use, the absence of meaningful recycling, and the continued enumeration of macro- and microplastics in sharks and other marine organisms, it does not require a crystal ball to assert confidently that research will expand considerably, as will the negative impacts.

Poetry

Shark-related poetry is a surprisingly rich genre that requires more of an aficionado and scholar than I to do it justice. Fortunately, the posting "Poems for Shark Week" exists on the poets.org website, which is a godsend for a poetry lover (yet dilletante) like your author.

Two sharky poems I knew before finding that website were written by American poet John Ciardi (1916–86). The first, a dark poem (the word *dark* appears three times in 140 words) entitled simply "The Shark," begins, "My dear, let me tell you about the shark, / Though his eyes are bright, his thought is dark," and ends, "That one dark thought he can never complete, / Of something—anything—somehow to eat, / Be careful where you swim, my sweet."

The second, more whimsical poem, "About the Teeth of Sharks," accurately describes the concept of serial tooth replacement in eight lines beginning, "The thing about a shark is—teeth, / One row above, one row beneath," and ending, "Now look in and . . . Look out! Oh my, / I'll never know now! Well, goodbye."

Poetry invoking sharks has been written by Walt Whitman, Carl Sandburg, James Dickey, Herman Melville, and numerous others. Do yourself a favor (once you have finished this book, of course), and visit the poets.org website devoted to this genre.

Finally, I count among my biggest honors my brief inclusion in a poem by Pulitzer Prize-winning Irish poet extraordinaire Paul Muldoon, very likely the first mention of a shark in one of his poems. The poem, which included references to other colleagues on a Semester at Sea educational cruise, was read by the author himself, and I was there to bask in the glory.

Polar Sharks

Polar seas are frigid, but that doesn't stop sharks from living in Arctic and Antarctic waters, broadly defined as those areas above latitudes of about 66° N or 66° S. However, only a select few sharks, about a dozen in Arctic waters and three in the Antarctic, possess the ability to maintain physiological function at low temperatures, and no shark species lives exclusively in polar waters.

For polar sharks and bony fishes, a major risk is the freezing of blood and other internal fluids. Some polar bony fishes have evolved the protective measure of producing antifreeze proteins that lower the freezing point of their internal fluids, but similar adaptations

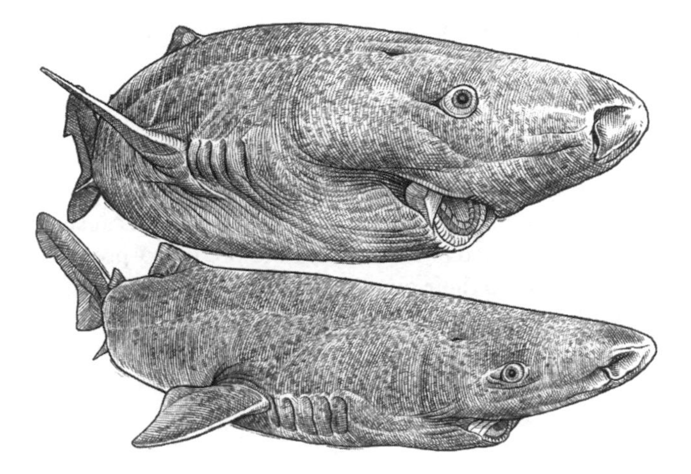

have not been identified in sharks, except for a single report of antifreeze proteins in the skin of an unidentified shark purchased from a supermarket—hardly definitive verification.

Whatever the underlying physiological mechanism that allows sharks to live in very cold polar waters, you should not be surprised that most polar sharks are quite sluggish, since metabolic rate correlates with body temperature, which is why, for example, iguanas fall out of trees when there is a cold spell in Florida.

Sharks of the Arctic include the Greenland Shark and Pacific Sleeper Shark, as well as the Porbeagle and Basking Shark. The Southern Sleeper Shark and Porbeagle also occur in the polar waters of the southern hemisphere.

Two final notes: Despite their reputations as the couch potatoes of the sea, Greenland Sharks forage on animals that might require bursts of speed to catch (e.g., seals, moose, and Polar Bears). And, second, living life in the very slow lane may confer longevity: one was recently estimated to be between 272 and 512 years old.

Predators of Sharks

Many shark enthusiasts are drawn to sharks because of the group's legendary and justifiable predatory prowess, but sharks have their own predators. Even the paragon of predatory skill, the White Shark, is not safe from becoming prey when in the presence of Orcas.

The best-known predators of sharks are other sharks. During a shoot for a NatGeo documentary, we set one

single hook using a small dead Finetooth Shark provided to us by other researchers after it had died. When we retrieved the line, we found two shark heads and a Bull Shark. A larger Sandbar Shark had taken the bait (the Finetooth Shark), only to become bait itself for an even larger Bull Shark, leading to my remark, which survived the cutting room floor and appeared on the documentary: "It's a shark-eat-shark-eat-shark world!"

So, larger sharks prey on smaller sharks. Other predators of sharks include birds, crocodiles, seals and sea lions, Orcas, dolphins, groupers and large bony fishes, and doubtless others.

Even though sharks eat sharks, the maxim that sharks won't eat other sharks is not entirely apocryphal. Many sharks will avoid eating rancid or otherwise rotting sharks, a fact that was utilized by the company Sharkdefense to develop a chemical shark repellent from decaying shark tissue.

Readings about Sharks

In Ernest Hemingway's 1952 masterpiece *The Old Man and the Sea*, he wrote of the Shortfin Mako: "Built to swim as fast as the fastest fish in the sea and everything about him was beautiful except his jaws . . . This was a fish built to feed on all the fishes in the sea, that were so fast and strong and well-armed that they had no other enemy."

The number of both novels and, especially, nonfiction books about sharks is an embarrassment of riches. Here are a few selections from the Western tradition.

In the realm of shark fiction not written for children, the most successful is obviously Peter Benchley's 1974 book *Jaws*. Benchley was so disturbed that his book had brainwashed generations with the stereotype of sharks as consummate killers, he spent the remainder of his life working for shark conservation. Most contemporary fiction that features sharks riffs on the same negative stereotypes, often with a twist: sharks being alien creatures, shark zombies, genetically modified super-beasts, megapredators unearthed by geological displacement of the sea floor, and so on. Fun reading, if that's your thing and you don't mind a large dose of gore, absurd situations, and horror, but not worthy of additional space here.

Most shark-related nonfiction books are field guides, beautifully illustrated coffee table books, personal narratives, natural histories, stories of shark attacks, and so on, written mainly for a general audience but useful to specialists as well. A sampling of these would include (in no particular order), *The Sharks of North America* (Castro and Peebles), *Sharks and Rays of Australia* (Last and Stevens), *Sharks of the World: A Complete Guide* (Ebert, Dando, and Fowler), *Sharks of the Shallows: Coastal Species in Florida and The Bahamas* (Carrier), *Sharks: The Animal Answer Guide* (Helfman and Burgess), *The Shark Chronicles: A Scientist Tracks the Consummate Predator* (Musick and McMillan), *Sharks of the World* (Compagno, Dando, and Fowler), *The Shark Handbook* (Skomal), *Demon Fish: Travels Through the Hidden World of Sharks* (Eilperin), *Sharks and People: Exploring Our Relationship with the Most Feared Fish in the Sea* (Peschak), *Shark Doc, Shark Lab: The Life and Work of Samuel Gruber* (Stafford-Deitssch), *Genie The Life & Recollections of Eugenie Clark* (Castro and Scholl), and, most recently, the books *Why Sharks Matter: A Deep*

Dive with the World's Most Misunderstood Predator (Shiffman) and *The Lives of Sharks* (Abel [me] and Grubbs). If you crave books on human-shark interactions, these may interest you: *Sharks in the Shallows: Attacks on the Carolina Coast* (Creswell), *Close to Shore: A True Story of Terror in an Age of Innocence* (Capuzzo), and *Twelve Days of Terror: A Definitive Investigation of the 1916 New Jersey Shark Attacks* (Fernicola).

Highly specialized nonfiction books on sharks are more technical and their content and writing may be accessible mainly to students and specialists. Some of these include: *Shark Biology and Conservation: Essentials for Educators, Students, and Enthusiasts* (Abel [me again] and Grubbs), *Sharks and Their Relatives II: Biodiversity, Adaptive Physiology, and Conservation* (Carrier, Musick, and Heithaus, eds.), and *Biology of Sharks and Their Relatives*, 3rd ed. (Carrier, Simpfendorfer, Heithaus, and Yopak, eds.).

If you read all of the books in this entry, or use them as references, you should be as well-versed on sharks as anyone.

Repellents and Warning Systems

During 2021 and 2022, my graduate student and I spent countless hours *unfishing* for sharks, that is, actively trying *not* to catch them using hook-and-line. Well, not really. What we, in fact, were doing was a rigorous and objective scientific project investigating technologies to discourage sharks from biting hooks. A baited hook that repelled sharks but not bony fishes could be of significant help in reducing shark bycatch on commercial longlines and even in recreational hook-and-line fishing.

As a model for a business, personal shark repellents for surfers and swimmers make an ideal product, a solution for which there is no problem, so to speak, since shark bites are extremely rare events. At the same time, however, repellents may reduce anxiety among users. Plus, given enough time, unlikely events of all stripes, including shark bites, often happen. Finally, if you are stranded at sea and endangered by one or more sharks, a repellent that buys you even a little time may be the difference between life and death.

Some shark repellents may be effective, but none are foolproof, and their efficacy depends on the application. Many of the larger scale repellents, such as mesh beach nets and culling, may unnecessarily kill sharks that are either harmless or are not endangering people. In South Africa, 27 mi (44 km) of beach nets caught an average of 1,470 sharks annually (in addition to 536 other large marine animals). Newer net materials and visual deterrent barriers that do not entrap sharks are being more widely tested or used.

In recent years, numerous personal deterrents employing different repellent techniques (physical screen, chemical, acoustic, visual or camouflage, electric, magnetic, chainmail wetsuits, and cages) have become available. Currently, most personal deterrents are proprietary and many have not undergone meaningful, robust scientific scrutiny that assesses both their efficacy and, ideally, their underlying physiological basis for repelling sharks.

Circling back to the testing done by my lab on hook technology, the preliminary results were not promising—that is, the hooks were not particularly effective in repelling the sharks that dominated our catches, Atlantic Sharpnose and juvenile Sandbar Sharks. Investigators studying similar repellent technology on other sharks had different results, and testing continues, including increasing the strength of the putative repellent.

Reproduction

If sex is of interest to you, it doesn't get any better than sharks' amazing array of reproductive styles and strategies. The diversity of reproductive specializations in sharks is nothing short of stunning, especially in the ways developing embryos are nourished. Some sharks lay eggs, while others are born live, but the different stories behind these two modes are simply mind-blowing.

Internal fertilization, which occurs in all sharks, results in offspring that are basically miniature to small adults, capable of fending for themselves from birth. Without parental care, which is absent among sharks, having a higher survival probability is critical. The number of offspring per female pregnancy in sharks ranges from one to more than three hundred.

If diversity is your passion, then you will be gobsmacked by the modes of embryonic nutrition in sharks and their relatives, of which there are about ten variations. Initial nourishment to all developing shark embryos is supplied from the yolk-sac, after which two basic patterns emerge. Either the embryos continue to receive nutrition exclusively from the nutrient-dense yolk produced by the mother and deposited in the egg (about 40 percent of sharks and rays), or the mother provides ongoing nutrition in other ways, including via a placenta or by bathing the embryo in uterine milk.

Let's consider a couple of the variations of embryonic nutrition. The most widely utilized type of live birth in elasmobranchs is yolk-sac viviparity, where the developing pup receives the bulk (but not all) of its nourishment from its yolk sac. A thin membrane, instead of a tough shell, is deposited around the developing eggs. About halfway through development, the embryos escape from this membrane and are then free in the uterus. Overall, about 40 percent of extant species of elasmobranchs practice yolk-sac viviparity, including all of the more than 160 species of sharks of the superorder Squalomorpha (thus, angel sharks, frilled sharks, saw sharks, dogfish, gulper sharks, sleeper sharks, lantern sharks, and cow sharks), plus some smoothhounds and wobbegongs.

In oophagy, which means "egg-eating," after most of the initial yolk has been absorbed by the embryos, the mother begins ovulating additional unfertilized ova, on which the embryos feed, producing protruding *egg stomachs*, the beer bellies of the shark set. Oophagy occurs only in lamniform sharks (fifteen species that include the five mackerel sharks, Crocodile Shark, and threshers) and the small family Pseudotriakidae (false catsharks).

Finally, monogamy is rare in sharks of both sexes. Also, in some species, sperm may be stored and used to fertilize multiple litters, leading to the phenomenon of *multiple parentage*—that is, the same or subsequent litters having more than one father.

Requiem Sharks

The requiem sharks, members of the family Carcharhinidae, include a wide variety of temperate and tropical sharks found in waters of continental shelves, estuaries,

coral reefs, open oceans to 2,600 ft (800 m), and even rivers. The origin of the word is thought less to connote a lament for the dead, as in a funeral dirge, and more of a derivative of the word *requin*, French for shark.

Requiem sharks include many familiar species, such as the Sandbar Shark, Blacktip Shark, Blacknose Shark, Spinner Shark, Blacktip Reef Shark, Silky Shark, Dusky Shark, Bull Shark, Grey Reef Shark, Caribbean Reef Shark, Silvertip Shark, Blue Shark, Copper (Bronze Whaler) Shark, Oceanic Whitetip Shark, and Whitetip Reef Shark, plus more than three dozen more.

The requiem sharks are the archetypical shark, with a long, arched mouth with blade-like teeth, which are often broader in the upper jaw, a nictitating membrane that protects the eye, and no spiracle. They are found in both inshore and pelagic waters from the surface to around 2,625 ft (800 m), where they feed on fish (including sharks), cephalopods, turtles, and mammals. They all give birth to live young, many are potentially dangerous, and several have been overfished. Over two dozen requiem shark species are in the range of 5–16 ft (1.7–5 m). Most are high-level mesopredators, and some coral reef requiem sharks are part-time apex predators.

Ridgeback Sharks

Ridgeback sharks are a non-taxonomic category of requiem sharks. The name refers to the eponymous slight but easily identifiable elevation of skin between the first and second dorsal fins on some members of this family.

Informally classifying sharks as either *ridgebacks* or *non-ridgebacks* was an attempt by US fishery biologists to impose some order on the often incorrect ways that both commercial and recreational fishers identified sharks so vulnerable species could be managed. There are also life history differences that conveniently are correlated with the presence or absence of the interdorsal ridge—for example, low reproductive potentials that led to long recovery times for some ridgebacks. The primary non-ridgeback shark species reproduce about twice as fast as the ridgebacks (a big exception is the Bull Shark, a non-ridgeback species whose life history is more in line with the ridgebacks).

Sandbar, Dusky, Bignose, Night, Silky, and Caribbean Reef Sharks are among the more prominent ridgebacks, while Bull, Blacktip, Blacknose, Finetooth, Spinner, Lemon, and Atlantic Sharpnose Sharks are non-ridgebacks.

In the US jurisdictional waters of the Atlantic, most ridgebacks are *prohibited* species, and must be immediately released if caught. Some non-ridgebacks are also prohibited in the Atlantic. Information on these can be found on the websites of state departments that deal with natural resources as well as those of the National Oceanic and Atmospheric Administration (NOAA).

Identifying sharks at sea and recalling which ones are prohibited are bewildering undertakings for many earnest fishers. For them, NOAA has a message: *if you don't know, let it go*.

Riverine Sharks (Freshwater Sharks)

Thought you were safe from sharks when swimming in your favorite river? Think again. Only four kinds of sharks are known to inhabit freshwater ecosystems, but these include species considered dangerous, like the Bull Shark. As a group, approximately 5 percent of sharks and rays enter brackish and fresh water and 3–4 percent (all rays) spend their entire lives in fresh water.

Why so few? First, the physiological systems of sharks are exquisitely tuned to the marine environments. While numerous sharks inhabited freshwater habitats during the Golden Age of Sharks, since then, sharks have evolved in marine ecosystems. The physiological and anatomical specializations required for sharks to inhabit freshwater ecosystems are so extreme that evolution did not favor them except in a few cases.

Fresh water could also impose constraints on the ability of sharks to detect the electrical impulses emitted by their prey. Sharks may also be somewhat disadvantaged energetically in fresh water, since the absence of salts in the water translates into less buoyancy, and thus a faster swimming speed is required to prevent the shark from sinking. Furthermore, freshwater environments may have limited niche spaces for additional shark species—in other words, *no vacancy*.

Besides the Bull Shark, the other three species of river sharks are all in the genus *Glyphis*, whose members

can be found in the western Indo-Pacific, Papua New Guinea, and northern Australia. All the riverine sharks have the physiological flexibility to live in the marine environment as well. None of the above species live and reproduce entirely in freshwater systems.

Some populations of batoids live predominantly in fresh water but can tolerate more saline waters, including the Atlantic Stingray from Florida's St. Johns River system, the Smooth Freshwater Stingray from Western Africa, the White-edge Freshwater Whipray from Southeast Asia, and the giant Mekong Freshwater Stingray, which is restricted to the Mekong and Chao Phraya Rivers in Laos and Thailand.

Almost an entire family of rays (approximately twenty-five kinds), the Potamotrygonidae, are landlocked freshwater species living in the Amazon basin. They are incapable of living in brackish or marine environments. The Largetooth Sawfish is also known to enter fresh water.

Except for the Bull Shark, all of the freshwater sharks and rays have relatively small regional distributions. Given that, and the environmental threats posed by development along rivers and so on, many populations are threatened.

Roadkill

Yes, *roadkill*, as in collisions with ships and propellers. The most susceptible to these impacts are the slow-moving planktivorous species of the ocean's surface, such as the Basking Shark and Whale Shark. But other species are often seen with propeller marks on their body.

In addition to being slow-moving, Whale Sharks and Basking Sharks are known to make long migrations, some of which take them across busy shipping channels. How widespread is the phenomenon? It is difficult to quantify, but it likely represents a not insignificant threat to some species in some locations.

Rogue Shark

Why do sharks bite people? Legendary shark biologist Samuel "Sonny" Gruber conjectured that bites might be due to entering a shark's space, perceived threat to or competition with the shark, interfering with a shark's courtship or mating, or simply some random event.

Some bites are doubtless related to feeding and hunger. For example, the numerous (often minor) shark bites that occur along beaches of the Southeastern United States every summer are attributed to Blacktip Sharks, swift-swimming, aquatic projectiles that probably confuse flapping human arms and legs for the small fish they eat, an understandable mistake in that area's murky water. White Sharks may confuse people for their pinniped (seal and sea lion) prey.

Some shark bites, like the above, are easily explained, or at least allow more educated conjectures than others, which defy easy, specific categorization. In a small number of cases, an additional explanation has been invoked, the idea of the *rogue shark*. Traced perhaps to the Jersey Shore attacks of 1916, or even much earlier, the notion refers to a single shark that wreaks often deadly havoc on the people in its path and has developed a taste for human flesh.

The major problem with this theory is the lack of evidence of multiple shark bites in a particular area over a short period. Yes, these have been reported (e.g., the 2010 series of three serious bites within minutes, and a fatality a few days later, at Sharm El Sheikh, Egypt). Even then, without specific knowledge that a single shark was responsible, alternate explanations that implicate different sharks are equally plausible. The scientific community requires solid evidence before accepting ideas like this one, and the evidence simply is not there.

Sandbar Shark

You may have guessed from the numerous mentions in *Sharkpedia* that, to me, the most iconic shark is the Sandbar Shark. Sandbar Sharks represent sharkdom in every aspect, including their commanding presence, ecological importance, conservation story, biological attributes, and finally, in no small part, their elegance.

We'll focus on their conservation story, after acknowledging their elegance. Whether watching Sandbar Sharks model their grace in the crystal-clear waters of an aquarium or seeing one emerge for a heartbeat from the murky depths of coastal waters, there is something special about this beast, with its enlarged first dorsal fin, docile nature, swimming loveliness, and slate gray-brown color palette. Yes, elegance incarnate.

Sandbar Sharks are found worldwide in temperate waters shallower than 330 ft (100 m), where they are primarily bottom-dwelling. Sandbars are among the most common sharks along the US East Coast and in Hawaii. They mature at about age fifteen and give birth to about eight pups every two years. They are considered mesopredators (one trophic level beneath apex predators) to small bony fishes and occasionally sharks, rays, crustaceans, and cephalopods (squid and octopuses).

Fifty years ago, Sandbar Sharks were abundant along the US East Coast, but beginning in the mid-1980s, they, along with many other sharks, began to be heavily fished both recreationally and commercially. By 1992, the main species driving the US shark fishery was the Sandbar Shark, in part because it was the most common large coastal shark from Florida to Virginia, but also because it had high-quality dorsal and pectoral fins greatly prized for the shark fin soup trade.

In 1993, the first fishery management plan for coastal sharks inhabiting US Atlantic and Gulf of Mexico waters was implemented. Sandbar Sharks were included in the "large coastal sharks" category, and the plan projected a ridiculously underestimated recovery time of only two years for the species. Sandbar Sharks have a very limited reproductive potential: a twenty-five-year replacement cycle (population doubling time), compared to about half that time for, say, Blacktip Sharks.

Currently, Sandbar Sharks in the US are still considered overfished, but overfishing is no longer occurring. Owing to their slow life history characteristics, however, Sandbar Shark stocks in the US are currently not expected to be rebuilt until around 2070. This assumes that human insults like climate change, coastal pollution, and so on do not derail recovery.

Save the Bay, Eat a Ray: Trophic Cascades

Trophic cascades are pretty neat ecological phenomena. Simplified, a trophic cascade refers to the dominating roles predators play in structuring ecological communities, and results from either adding or removing predators from the trophic pyramid. When the latter happens—say, by overfishing—the overfished species' prey, now lacking the controlling presence of their predator, may explode in numbers, a phenomenon called *trophic* (or *predator*) *release*. The repercussions may resonate even further down the food chain, since the newly expanded trophic level may deplete the prey lower in the food chain.

Examples of trophic cascades abound, including the classic (if oversimplified) case of the disappearance of sea otters along the California coast, leading to increases in their prey (sea urchins) and the subsequent drastic loss of kelp forests (the food of the sea urchins).

You might think that numerous cases of trophic cascades involving sharks would have been documented, but that has not been the case. Further, one highly publicized case of a trophic cascade in the shark world has been challenged as inaccurate.

A 2007 paper in the journal *Science* titled "Cascading Effects of the Loss of Apex Predatory Sharks from a Coastal Ocean" claimed that population decreases of large coastal sharks (e.g., Sandbar, Bull, and Tiger Sharks) along the US East Coast triggered a trophic cascade in which the populations of their prey (e.g., Atlantic Sharpnose Sharks and Cownose Rays) then swelled.

In the case of the Cownose Ray, the study asserted that the apparent population rise caused a population crash of the ray's chief prey, the Bay Scallop, and the subsequent loss of the North Carolina fishery on which it was based.

A series of rebuttal papers used sound scientific evidence to challenge this conclusion. For example, the cause of the collapse of the scallop fishery in North Carolina was best explained by a combination of overfishing, loss of seagrass habitat, hurricanes, red tides, wasting disease, and pollution, as well as a poorly regulated recreational fishery. Scallops were not even the ray's prey of choice.

The initial report contributed to a large-scale but wrongheaded movement to cull the US Atlantic population of Cownose Rays, an elasmobranch with one of the lowest reproductive potentials known. In the Chesapeake Bay, the "Save the Bay, Eat a Ray" campaign started, and many well-meaning restaurants unwittingly promoted eating the animal least capable of supporting a fishery.

In 2019, the species was listed by the IUCN as "vulnerable," and since then contests designed to kill large numbers of Cownose Rays have been discontinued, but only in Maryland, as the initial misinformation lives on. This case of well-intentioned if unsound science represents one of the shark and ray world's sadder conservation stories.

Senses

In many areas, if you're emersed in coastal waters, chances are a shark has sensed your presence. An earlier entry established that sharks have an acute sense of smell, but their other senses are equally prodigious, and they play no small role in the predatory prowess of sharks.

Senses allow sharks to interpret their surroundings and respond appropriately. These include senses that humans and other vertebrates possess, as well as one totally foreign to us. A shark's sensors enable it to detect prey, predators, conspecifics (members of the same species), other organisms, and obstacles, but also allow a shark to orient itself.

Sharks and rays possess photoreceptors (vision), chemoreceptors (smell and taste), mechanoreceptors (touch and hearing), electroreceptors (detecting an electrical field), and possibly also magnetoreceptors. We'll consider only hearing and electroreception here.

Yes, sharks have ears, though you would see them only upon dissection, the only external sign being small, paired *endolymphatic pores* on the top of the head. In 2015, a rescue helicopter hovered over a shipwreck survivor in the water near Aruba, and was about to pluck him from the water when a shark bit and killed him. Immediately, speculation surfaced that the low-frequency *whomp-whomp-whomp* of the helicopter's rotors played a role in attracting the sharks. Was that possible? Yes. While the shark that killed the shipwreck victim could have been attracted using any of its suite of senses, or it could have coincidentally encountered the victim, the sounds of the hovering helicopter could have played a role, since low-frequency sounds attract sharks, although irregular sounds are better attractants.

Although the entire suite of senses integrates with each other to paint a complete sensory picture of the environment, different senses come into play at various distances and for specific functions. These are (from farthest to nearest): hearing, smell, vision and lateral line (current, vibration, and pressure sensors), electroreception, and touch and taste.

Once a shark has located its prey, it is guided over the last approximately 1.6 ft (0.5 m) to its destination by receptors called *ampullae of Lorenzini*, concentrated on the head. These are capable of detecting extremely weak electric fields, such as those emitted by a prey item (and indeed all animals), especially one that might be bleeding. That close to a prey item, the sensory cues that guided the shark in from afar may now fail it. The pioneering studies that illuminated the fact that sharks possessed an electrosensory system were conducted by Adrianus Kalmijn of Scripps Institution of Oceanography.

You should be comforted (unless you happened to be the unfortunate shipwreck victim discussed above) that sharks have detected your presence when swimming at the shore and have chosen to ignore rather than menace you.

Serial Tooth Replacement

Information on the early evolution of sharks and their cartilaginous relatives is scarce, since cartilage in general does not fossilize well. Ironically, however, among the most numerous vertebrate fossils are shark teeth, which are more heavily mineralized and thus endure. Thus, paleontologists know a lot about shark dentition, and this information has been useful. For example, teeth from ancestors of the cowsharks have been discovered

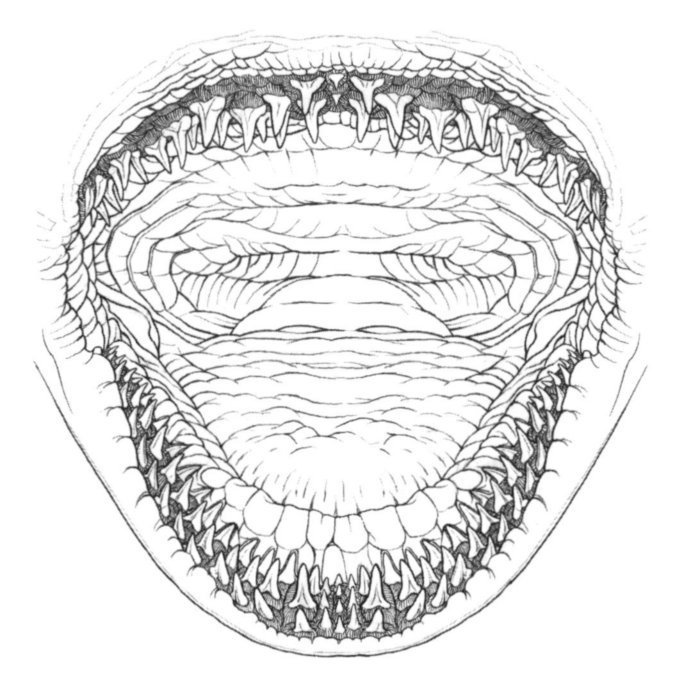

in sediment 180 to 200 million years old, establishing these species as the most primitive sharks.

These discoveries are possible because of the phenomenon of serial tooth replacement, in which multiple rows of teeth in various stages of development are anchored somewhat loosely in collagen, a fibrous protein, and are not embedded in the cartilage itself. The collagen advances in conveyor-belt fashion and ensures that one or more rows of teeth are always available and functional. The process of replacing a row takes only several weeks, and varies among sharks.

It may seem wasteful to produce and then shed teeth continually, since resources such as minerals and energy must be allocated to producing teeth, meaning that these are diverted from other uses. For predatory vertebrates with a permanent set of teeth, losing one—especially a canine tooth—might doom it to starvation, a major cause of death among top predators. When a 10 ft (3 m) Lemon Shark leaves a couple of teeth embedded in the wood rub rail on the bow of our research vessel after biting it (barely missing my foot, as it were), replacement teeth would already be in line.

Shark Burn

Enlightened people would seem to want to avoid getting bitten by a shark. Many of my students would beg to differ. On one occasion, when a student suffered a minor bite from a young juvenile Lemon Shark, I overheard other students groaning that they wish it had been them.

Which brings me to *shark burn*. You might be tempted to conclude that the term comes from the US television show *Shark Tank*, where millionaire venture capitalists called "sharks" savage the ideas of entrepreneurs who might rightly claim to be victims of shark burn. Or it could refer to the feeling that you have been taken advantage of by members of some professions who have been referred to as *sharks*.

In reality, shark burn refers to skin abrasions on people who handle sharks and, in the process, rub against them, typically against the grain of the shark skin—that is, from back to front. Shortly after the abrasion occurs, a rash develops, similar to rug burn. The rearward-facing dermal denticles of the shark are the culprit, and the degree of shark burn that one might experience differs by species.

How does this relate to wanting to be bitten by a shark? Dermal denticles are built on the same structure as teeth. Thus, if you experience shark burn, claiming that you survived being bitten by a shark is not entirely disingenuous. My students seem to agree, since after working with a shark before releasing it, they often proudly display their shark burn.

Shark Camp

This entry is not about a summer interlude for adventure-seeking kids, but rather what the Merriam-Webster dictionary defines as "something so outrageously artificial, affected, inappropriate, or out-of-date as to be considered amusing." Camp, according to that definition, abounds in the shark world, in movies, comics, documentaries, and so on.

Consider shark movies, whose campy titles include *Santa Jaws*, *Atomic Shark*, *Toxic Shark*, *Sharkenstein*, *Ice Sharks*, *Ozark Sharks*, and *Sky Sharks*. These may be representative of the genre, but the apex of shark camp belongs to the *Sharknado* franchise, of which there are apparently an original and five sequels (I've seen none of them). In these cult flicks, hungry sharks are entrained in a waterspout and then are plunked down where they terrorize a cast of unknown actors or aging celebrities trying to start or resurrect a career. Some films that also pass themselves off as documentaries are, in fact, camp. Paramount among these is *Megalodon: The Monster Shark Lives*. It is difficult enough to educate the public about believable but incorrect assumptions or conclusions about sharks without the tidal wave of pseudoscience that some shark documentaries purvey.

Shark-thamology

If a 15 ft (4.5 m) Tiger Shark swims by you and raises its white nictitating membrane, as happened to one of my colleagues, either the shark thinks it is endangered by you, or you are being vetted for inclusion on its menu. Nictitating membranes are scale-covered protective eyelids found in the requiem sharks (plus some birds,

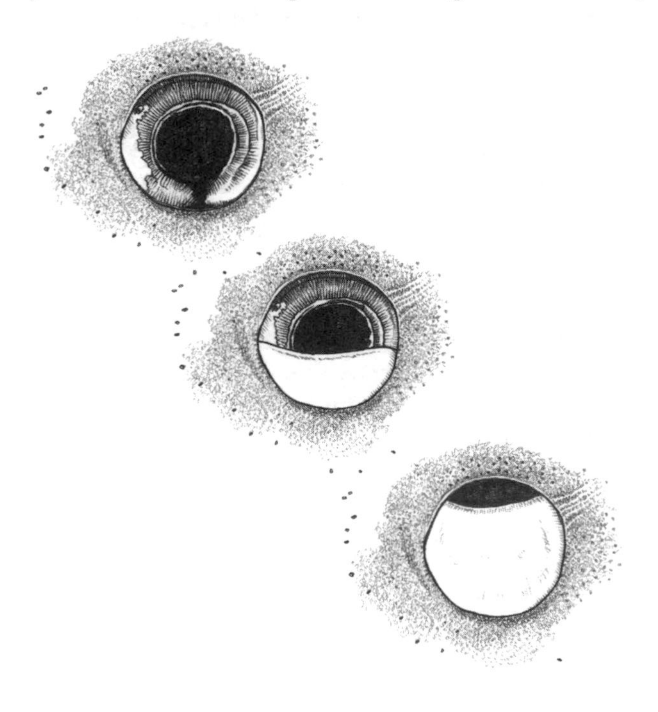

lizards, frogs, seals, Polar Bears, camels, and Aardvarks). They are deployed as the shark is about to eat, or when it is threatened.

There are at least three other ways in which sharks protect their eyes. White Sharks roll their eyes 180° in the socket, exposing the whites. The cowsharks and many relatives in their superorder retract their eyes into their sockets. Finally, Whale Sharks were recently found to possess denticles on the surface of their eyes.

Shark Week/Sharkfest

After appearing in several shark "documentaries" on the NatGeo cable channel, I vowed never to do another. Many of my colleagues have made similar decisions.

Discovery Channel's Shark Week and National Geographic's Sharkfest saturate cable airways in over seventy countries for a week or more each summer and in countless replays, with recent annual viewership exceeding fifty million.

Giving the devil its due, these shows have elevated shark appreciation. At the same time, however, many of these shows do the opposite in their attempt to entertain more so than educate. The worst of these validate irrational fears, provide dubious facts, may be totally fictional, misuse science, set up ridiculous scenarios, and waste resources, including viewers' time.

I suspected these drawbacks before agreeing to do my first show. But I believed that I could manage my participation, specifically discussing only issues that were scientifically accurate and not beyond my expertise. And, in most of my appearances, this was true. But I was a babe in the woods and I may have unwittingly contributed to some of the negatives of the genre. You may wonder why I agreed to participating in so many; I do as well.

More recent iterations on cable TV have included honest-to-goodness real science amidst some of the silliness. Stay tuned.

Sibling Rivalry/Embryophagy

With only 541 named species, sharks are a marginally biodiverse group. But in terms of how their embryos are nourished, their biodiversity surpasses that of all other vertebrates. Among these, one is so unimaginable that it merits its own entry here, *intrauterine cannibalism*, also known as *embryophagy* or *adelphophagy*.

Yes *cannibalism*, while still an embryo inside its mother's uterus! Though not unique (it has been documented or strongly suspected in other sharks), this bizarre adaptation is best known from the Sand Tiger. Within each uterus, the fastest-growing or oldest embryo assumes the role of apex predator and methodically stalks and consumes its smaller siblings. In the embryos of most shark species, teeth are among the last major anatomical structures to develop and they do not break through the gums until slightly before birth. Sand Tiger embryos, however, exhibit precocious tooth development. At a size of about 5 in (12 cm), the largest embryo starts eating as many as a dozen other embryos and stores the energy in its yolk stomach. At birth, the two newborns are each greater than 3.3 ft (1 m). Although no longer apex predators in the larger ocean into which they were born, the newborn Sand Tigers are formidable predators in their own right.

Stock Assessments

As the name denotes, stock assessments are studies that elucidate the health of shark stocks. This sounds much easier in theory than it is in practice.

In order for a stock assessment to be useful to policymakers, marine ecologists, and fishery managers, it would need to be comprehensive, accurate, recent, and reliable. The first consideration is who collects the data. Typically, information on stocks comes either from recreational and commercial fishers themselves (*fishery-dependent* data) or from marine scientists (*fishery-independent* data). Even if the fishery-dependent data are earnestly collected, fishery-independent surveys are more trustworthy, and are often critical to meaningful stock assessments. For example, whereas fisheries often use methods that catch only specific portions of the populations of sharks, fishery-independent surveys can be designed to catch sharks that represent the entire population (e.g., by using multiple hook sizes). Moreover, many commercial fishers have trouble identifying sharks, and especially distinguishing between species that closely resemble each other.

To manage a shark, or indeed any fishery, having complete information about the animal's life history is vital. This includes size and age at maturity, size and age at first reproduction, fecundity (both per pregnancy and lifetime), frequency of reproduction, survival of offspring, and maximum size and age.

Because stock assessments are expensive, time-consuming, require expertise, and must be reconducted periodically, they have not been done for most species of sharks and batoids, which are then said to be *data deficient*.

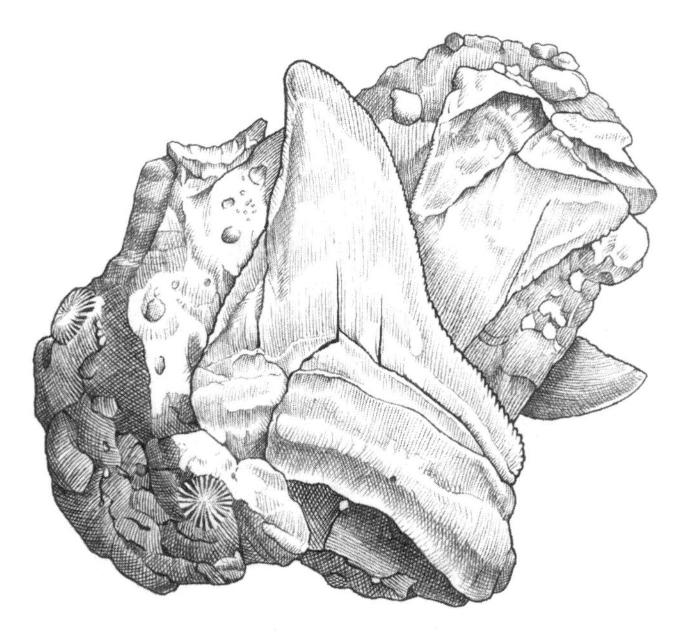

Stone Tongues of Dragons

The field of paleontology owes its early development to a shark and the Danish anatomist Nicholas Steno. Steno was one of the original myth-busters, and his powers of scientific observation were pivotal in developing the field of paleontology, but it required his dissecting a dead shark in the year 1666.

Then, dissections were not college biology laboratory activities that could be completed in three hours. An early anatomist could spend weeks or months of full days teasing out anatomical features, and the results were often works of art. During one such dissection, Steno connected the shape of the shark's teeth to tongue stones, or *glossopetrae*, thought to be rocks or the stone tongues of dragons.

We now realize that tongue stones were one of the most abundant vertebrate fossils: shark teeth. But in 1666, the process of fossilization was unknown, and what we call *fossils* were thought, among other incorrect explanations, to have dropped from the heavens. Steno not only concluded that the stone tongues were from sharks of the distant past, but he also provided a mechanism for their conversion to stone and he further hypothesized how the fossils wound up in rocks.

Studying Sharks

Unraveling the ecological and physiological mysteries of sharks involves different challenges than studying terrestrial predators like the big cats. For one, we are but clumsy interlopers in the underwater environment of sharks (although, truth be told, I am a clumsy interloper in most terrestrial ecosystems). Also, while diving with sharks is relatively inexpensive, ships and equipment are not, especially when studying deep-sea and pelagic species.

Handling sharks often presents a new suite of problems. Think sharks with large teeth that they are not afraid to use, unhappy to find themselves unwittingly plunked on a culling table. I know many colleagues who have suffered serious injuries from shark bites (and I have had several close calls and non-serious bites), or even stinging shark tail slaps (two of which launched pretty decent watches off of my wrist and into the briny).

A final challenge to studying sharks lies in achieving the meaningful sample sizes needed to validate the significance of findings. This is one reason why model organisms such as fruit flies, zebra fish, and white mice are used in scientific experiments.

Tagging sharks will be discussed in a separate entry. An alternative—camera traps—enables real-time surveys. My colleague Erin Burge uses the shark cam at Frying Pan Tower in North Carolina. A random visit to the shark cam's YouTube revealed a couple of big Sand Tigers, and he has spotted White Sharks as well. Studies using baited remote underwater video (BRUV) surveys have also become common.

One method beginning to be more widely used is based on the assumption that every living creature leaves fragments of its DNA (e.g., skin cells, fecal matter, mucous, blood) in the water after it has moved on. Analyzing water samples for this eDNA (environmental DNA) can allow scientists to assess if a species of shark was present in the area. Cool as this is, the technique is fraught with drawbacks. It may not detect a species that is present, and it provides no information (at least currently) on how many individuals are present, sex, size, reproductive status, or trophic status.

Tagging

On our cruises, one of our most gratifying occurrences is recapturing a shark that we had previously tagged. Consider the odds against this. First, the shark had to survive the capture process, which may involve significant physiological stress. Then, once released, the shark may require additional time to recover, during which it may be vulnerable to predation, or it may even succumb from the stress. Thereafter, the shark, which has a good chance of being highly migratory, must navigate what may seem like an ocean minefield of recreational and commercial fishing hooks and nets, polluted waters, and a sea full of potential predators. If it is lucky enough to survive these, *and* the shark returns to where it was tagged by us at one of our standard sampling locations, *and* the shark is hungry, *and* it is one of our sampling days, *and* it locates the hook, *and* another shark does not eat it before we retrieve the longline (a phenomenon called *depredation*), *and* the tag has not fallen out, we will have been lucky enough to collect the valuable data from the shark.

We have tagged Sandbar Sharks, Blacktip Sharks, Finetooth Sharks, Atlantic Sharpnose Sharks, Bull Sharks, Lemon Sharks, Sand Tigers, Great Hammerheads, Blacknose Sharks, Spinner Sharks, Bonnetheads, and a few others. For conventional tags like some of ours that require recapture, our recapture percentage is low, below 2 percent. In one study, we tagged 407 neonate (newborn) Atlantic Sharpnose pups, of which ten were recaptured, all within twenty-eight days, by us and by recreational fishers who reported their recaptures—a rate of 2.4 percent. We were surprised by the high recapture rate, considering the diminutive size of the tagged sharks (as small as 8 in or 20 cm) and the heavy predation on them by a variety of sharks, larger bony fishes, dolphins, and shorebirds. One was recaptured at the same location an hour after it was tagged, and another was recaptured 17 mi (27 km) away seventeen days later. From this tagging study we learned about the growth and movements of this important prey species.

Determining how fast sharks grow and where sharks move are central to understanding their ecology, and tagging studies provide insight into these. An array of tagging and telemetry tools, ranging from the very simple to the complex, are currently used for these purposes. Modern tags include: *Conventional tags*, like those described above, which require recapture; *PIT tags* (Passive Integrated Transponders), rice-grain sized tags like the microchips implanted in dogs and cats, or a car's E-ZPass; *Acoustic tags*, typically surgically inserted into the abdomen, which emit an aural ping and require a receiver nearby to detect; and *Satellite tags*, expensive game-changers that transmit radio signals to satellites whenever the tag's antenna is above the water surface. Satellite tags cost $1,000–$5,000 USD and the satellite service may cost an additional $500 USD per tag.

Threats

In the long term, sharks are no more imperiled than any of creation, given a growing human population and its juggernaut tendencies to overexploit resources and destroy nature, especially the dual insults of climate change and biodiversity loss.

In the short-term, there is both good and bad news about the status of shark populations. The good news is that more people and governments are beginning to appreciate the intrinsic, ecological, and even financial value of sharks in their habitats more than as food or products. This has led to effective, science-based management of some species in a few countries, widespread domestic and international prohibitions on taking many species, designation of some areas (e.g., The Bahamas and Palau) as shark "sanctuaries" (where some kinds of fishing are banned), and finning bans across the globe.

But there is an abundance of bad news. First, fishery managers lack the information about the life history characteristics and fishing mortality for many sharks and rays, which can hinder efforts to conserve them. Nor do scientists understand the health, behavioral, and ecological impacts on sharks of many environmental threats. What we do know is that removing them from an ecosystem causes changes in the ecosystem, although accurately understanding those changes is not easy. But an ecosystem with a healthy shark population is an ecologically healthier place than one without its sharks.

Overfishing, through both intentional and incidental catches, is overwhelmingly the principal threat for sharks, followed by habitat degradation and destruction, pollution, and aggregated human disturbance (persecution, noise pollution, etc.). A 2021 global Red List analysis found that almost all chondrichthyans (99.6 percent) are affected by fishing, and that overfishing is the main risk for all (100 percent) of the 391 threatened chondrichthyans. About one-third of threatened species are also imperiled by habitat degradation, primarily resulting from development (25.8 percent) and agriculture or aquaculture (9.5 percent). Pollution is a key risk for 6.9 percent. Climate change currently affects 10.2 percent of threatened chondrichthyan species through the degradation of coral reefs and/or ranges shifting toward the poles as waters warm. Among sharks, apex predators are particularly vulnerable to many of these threats because they are the least abundant organisms in an ecosystem, and they may concentrate and magnify levels of some toxicants.

What are some of the impacts of climate change on sharks? The impact of ocean acidification has the potential to be enormous in scale, causing changes in ocean chemistry that have not been seen in 65 million years, which will affect the vitality and survival of all taxonomic groups, sharks and shark food alike. And temperature change will have both physiological and ecological impacts as well. Consider the latter; many sharks and rays would be forced to migrate to higher latitudes or deeper water as temperature increases. For example, juvenile Bull Sharks have been found in North Carolina estuaries that had not previously been a frequently used habitat, perhaps because of the early arrival of summer temperatures. In moving to higher latitudes, sharks and rays may encounter ecosystems novel to them. These may cause problems like changes in the abundance and size of shark populations; changes in the food chain; alterations in behavior; and possibly mortalities, extirpations, and even extinctions of species for which migrations may be difficult or improbable or that may already be depleted or threatened by other stressors.

Tiger Shark

White Sharks may display more spectacular behaviors, like breaching, but an equally iconic species is the Tiger Shark. Tiger Sharks are found worldwide in tropical and temperate coastal waters, where they may reach 16.5 ft (5 m). This species is easily identified by its markings, which are most vivid in juveniles, its long caudal fin, and its wide, multicusped teeth. Surprisingly, Tiger Sharks have relatively weak jaws that bend across the body of large prey (e.g., sea turtles and dead whales).

They then twist or spin their bodies to carve out huge chunks of flesh. Tiger Sharks may produce sixty or more pups every three years, and their populations are considered healthy globally because of this high reproductive potential. However, a recent study showed a 71 percent decline in Tiger Sharks along the east coast of Australia.

Tonic Immobility

"Tonic immobility" does not refer to when my car won't start and I am unable to drive to the supermarket to get one of my wife's favorite beverages. Rather, it is a fascinating adaptation also known as *death feigning* or *thanatosis*, that many animals employ. The phenomenon has been observed in insects (e.g., the Blue Death Feigning Beetle) and all major extant vertebrate groups. Humans are often advised to adopt the strategy when an aggressive Brown Bear is nearby. In sharks, tonic immobility is almost always discussed in the context of being induced rather than as a part of their normal behavioral repertory.

The adaptive value of death feigning is most often as an anti-predation strategy, in that a predator may consider the prey as having been dead for a long while, and thus potentially toxic. Or the predator may be programmed to catch only moving prey. Tonic immobility may also prevent harassment by members of the opposite sex during courtship and mating.

Tonic immobility can be induced in a number of shark species, a growing list that includes Tiger Sharks, Lemon Sharks, and Whitetip Reef Sharks. Typically, in these scenarios a captive shark is flipped on its back, after which it visibly calms, typically within seconds. Scientists take advantage of this when studying live sharks that otherwise might be difficult to handle or that might struggle in captivity at levels that could induce dangerous stress. Sharks typically emerge from short bouts of tonic immobility when they are righted and, in some cases, induced to move.

The physiological basis of tonic immobility has also been studied in juvenile Lemon Sharks and other species. The physiology seems to be complex, and may involve stimulation (or overstimulation) of the parasympathetic nervous system, the so-called *resting and digesting* part of the nervous system that is calming and slows body functions (as opposed to the sympathetic nervous system, which is responsible for the fight-or-flight response).

These observations on tonic immobility in sharks shed little light on whether the behavior has adaptive value in their daily lives or, like the human appendix, may be an evolutionary holdover whose function is enigmatic.

Tropical Sharks

There may be no more ideal place to see sharks than the tropics, areas north and south of the equator where the minimum water temperature does not drop below 68° F (20° C), roughly corresponding to latitudes between the Tropic of Cancer (23.5° N) and Tropic of Capricorn (23.5° S). This expansive area includes ecosystems such as coral reefs, mangrove systems, and seagrass beds.

More than two hundred shark species and 250 kinds of batoids (skates and rays) inhabit shallow tropical ecosystems. About half of these are found exclusively in the tropics. Among the tropical sharks, the most speciose (species-rich) taxa are the catsharks (three families), the houndsharks (Triakidae), and the requiem sharks (Carcharhinidae).

If ecosystems can be labeled iconic, coral reefs would be at the top of the list. They cover only about 0.2 percent of the ocean floor, or 96,000 square miles (250,000 square kilometers), an area roughly the size of the state of Texas. But they represent some of the most important real estate on the planet. The major sharks on Indo-Pacific coral reefs include the Grey Reef Shark, Whitetip Reef Shark, Blacktip Reef Shark, Tiger Shark, Silvertip Shark, Galapagos Shark, Scalloped Hammerhead, Great Hammerhead, Blacktip Shark, wobbegongs, Tawny Nurse Shark, Sicklefin Lemon Shark, and Epaulette Shark. There is a lower diversity of Sharks on Atlantic reefs. Representatives include the Caribbean Reef Shark, Tiger Shark, Blacktip Shark, Blacknose Shark, Atlantic Sharpnose Shark, Lemon Shark, Bull Shark, Nurse Shark, and Great Hammerhead.

Mangroves, one of the most biologically productive ecosystems on the planet, are a broad group of salt-tolerant trees that grow at the water's edge in the tropics. They play important roles in the life histories of numerous sharks as well as critically endangered rays, like the sawfish. Mangroves provide myriad hiding places for a great diversity of marine life, especially food for sharks. They also provide nursery habitat for some shark species, like Lemon Sharks in Bimini, The Bahamas.

Mangrove communities are at high risk from development, especially in countries without land-use controls.

Mangrove environments are cleared for aquaculture (especially shrimp), resort developments, and housing. They are also threatened by rapid sea-level rise, agricultural runoff, deforestation for biomass fuel, and oil and gas exploration and production. Already, as much as 50 percent of the world's mangroves have been lost, and more have been degraded by pollution.

Twentieth Century Shark Biologists

Here is only a small sampling of influential twentieth-century shark biologists. Read Jose Castro's outstanding 2017 article "The Origins and Rise of Shark Biology in the 20th Century" for a more thorough treatment.

I selected the following people based on my worldview as a shark biologist formally educated in the United States. That means my list is woefully incomplete; it does not mean that the people I selected are unworthy of inclusion here or that others whom I omitted did not deserve mention. It also reflects the lack of opportunities for many. When an entry for twenty-first-century shark scientists is written, it will doubtless be far more diverse!

Let's do these alphabetically (so says your author, who typically is near the top of alphabetical lists). Aristotle (384–322 BCE), you may notice, is not a twentieth-century scientist, nor was he the first to make significant observations about sharks. But he was an insightful scientist who, to name only a sample of his contributions, wrote about the reproductive physiology of sharks and rays, recognized that some sharks gave birth to live young, described the mermaid's purse egg cases of skates and some sharks, observed that some sharks utilize nursery grounds, and identified claspers as having a reproductive function (although not the one he conceived, as we have already established in another entry).

Among John "Jack" Casey's contributions as a fishery scientist at the US Fish and Wildlife Service was the initiation of the Cooperative Shark Tagging Program in 1962, a model for citizen-science that is still ongoing under the auspices of the National Oceanic and Atmospheric Administration (NOAA). Casey was also a pioneering shark scientist with contributions in the areas of age and growth, reproduction, and foraging.

On May 4, 2022, the US Postal Service honored Eugenie Clark (1922–2015) with a commemorative stamp. Known as the Shark Lady, Clark remains one of the most prolific (with more than 175 scientific articles and two books) and distinguished shark researchers of the twentieth century, at a time (post–World War II) when few females were working as scientists. Her accomplishments include the discovery of sleeping sharks and the fact that the Moses Sole synthesizes its own shark repellent, as well as establishing that sharks are capable of cognition.

Leonard Compagno could be called a shark taxonomist's taxonomist, and has been considered the world's foremost shark expert. He has written over a hundred scientific papers, but he is best known for detailed books on shark identification and taxonomy (especially his 1984 catalog *Sharks of the World*, published by the Food and Agricultural Organization of the United Nations).

You don't get more iconic than Jacques Yves Cousteau (1910–97), whose pioneering documentaries and books inspired generations to choose marine biology as a career. Sitting prominently on my bookshelf is his 1970 book, coauthored with his son Philippe, *The Shark: Splendid Savage of the Sea*. Three of his documentaries focused on sharks, especially shark behavior. As coinventor of the Aqua-Lung, Cousteau's impact on marine biology as a whole is outsized.

Perry Gilbert (1912–2000) was the prominent shark behaviorist of the twentieth century, although his contributions to shark science were broader and included studies on vision, chemoreception, bite force, and shark anatomy. He wrote about 150 scientific papers as well as two books. Perry codeveloped the important and widely used fish anesthetic MS-222 (tricaine methane sulphonate, a chemical cousin of cocaine, a similarity that got me detained by authorities once when I crossed international boundaries with small packets of the white powdery MS-222 crystals for my research).

The next two people on my list I count as mentors, colleagues, and friends. Jeffrey B. Graham (1941–2011) was my advisor for my dissertation work on heart function in sharks in the Physiological Research Laboratory at Scripps Institution of Oceanography. One of the most intuitive and insightful biologists I have ever known, Jeff studied organisms living at the edge (i.e., those living lifestyles unlike their close relatives, such as air-breathing fishes, snakes that live in the sea, and warm-bodied bony fishes and sharks). We worked together with Ralph Shabetai, a human heart expert, on unraveling the mysteries of heart function in sharks.

Samuel "Sonny" Gruber (1938–2019), more familiarly known as "Doc," could also lay claim to being the world's foremost expert on sharks. I know Doc from my nearly thirty-year affiliation with the Bimini Biological Field Station, which he established in 1990 and where I have been taking my Coastal Carolina University students for an annual Biology of Sharks course. Doc's contributions to the knowledge of sharks, and to educating students and the public about them, were immense. He published about 190 scientific papers and supervised a generation of graduate students who have become a who's who of prominent shark behaviorists, ecologists, physiologists, and educators.

During the years we overlapped at Scripps Institution of Oceanography, I will always regret that I never crossed paths with Adrianus Kalmijn (1933–2021). Like others in this entry, Kalmijn was a brilliant scientist, and his work has been called monumental. His focus was shark sensory biology, particularly electromagnetic reception. His early work demonstrated that electric fields with exceedingly low voltage gradients of 0.01 microvolts per centimeter elicited a physiological response in sharks, and that the ampullae of Lorenzini were the electroreceptors.

To hear some of his colleagues discuss Don Nelson might lead you to think he merited the scientific equivalent of sainthood. He is best known for his understanding of agonistic displays (threat postures) in Grey Reef Sharks at Eniwetok Atoll. This work was important in not only elucidating the heretofore unknown behavior, but also in determining that shark aggression could be due to perceived threat and not necessarily predatory behavior.

Space permitting, entries for the following would have been justified: Barbara Block, Frank Carey, Jeff Carrier, Jose Castro, Jack Musick, Art Myrberg, Wes

Platt, Richard Rosenblatt, Stewart Springer, and, I'm sure, numerous others.

UREA-LY Want to Know How Sharks Taste?

In one of my other books, *Shark Biology and Conservation*, a subject heading reads "How Do You Solve a Problem Like Urea?"—a pun based on the song "How Do You Solve a Problem Like Maria?" from the iconic movie *The Sound of Music*. Since the pun understandably fell like a lead balloon on many younger readers, a new pun was called for, hence *urea-ly*.

There is reason to this punning; the chemical compound urea is central to how a shark achieves the critical water and salt balance its body requires, and it explains why much shark meat may be unpalatable. If water and salt levels are out of balance, the physiological processes on which life depends—for example, conducting nerve impulses or the beating of the heart—malfunction, with often fatal consequences.

Aquatic organisms—freshwater, marine, and in between—are faced with a fundamental problem: both the chemical composition and chemical concentration (the saltiness) of their internal fluid environment differ from those of their external environment, conditions that the physical laws of diffusion and osmosis try to correct. If allowed to do so, physiological chaos results. To counteract these, organisms must expend energy.

For a complete explanation of how sharks achieve water and salt balance, see my other book, *Shark Biology and Conservation*. Basically, it boils down to this: sharks maintain the balance by elevating their internal concentration of urea, a metabolic waste product.

Now, about the taste. Urea makes the meat of many sharks taste bad, especially if it is converted to ammonia. The solution, which I learned in my shark-eating days as a graduate student (it's been decades since I have eaten shark), was to marinate the shark before cooking to remove the offending chemicals.

Walking Sharks

Sharks have various modes of transportation. We have already discussed two, swimming and jumping. To this we add *walking*, which occurs in several species of epaulette and bamboo sharks, small (most

less than 1.6 ft or 0.5 m), sluggish, tropical Indo-Pacific sharks. Some epaulette sharks inhabit isolated, shallow tidepool habitats. These offer many advantages, especially abundant prey and protection from predation. But the prey in some tidepools may not be abundant, or the water can evaporate, or environmental conditions may be poor, such as temperatures that may exceed 93° F (34° C) in the scorching sun, much higher than in the nearby deeper reef. Not to worry! The group has adaptations that allow it to crawl, using its pectoral and pelvic fins, and wriggle, not unlike a salamander, over jagged, emergent coral to another, more hospitable tidepool.

Warm-Bodied Sharks

I often ask my students sitting in classrooms whose temperature is comfortably cool, say, 70° F (21° C), how long they could tolerate that temperature. For most, the answer is almost always *indefinitely*. Then I ask, "What if you were in water of the same temperature?" The students intuitively know that the answer is *much less*. I then follow with, "But 70° F (21° C) is the same temperature, in both water and air, no?" This often stumps them. But the point is salient: the medium is the issue, and it is really, really difficult for any warm-blooded, or endothermic, animal to retain its heat in water.

Water is a thief of heat. In the above scenario, hypothermia in water sets in quickly, and in two to seven hours a person becomes exhausted, may lose consciousness, and may even die, at an air temperature that would be tolerable.

It is thus understandable that most aquatic vertebrates are ectotherms—that is, their body temperature is essentially the same as their environment's. Exceptions are aquatic mammals and birds, plus a small handful of bony fishes and elasmobranchs, which trap the heat that their bodies produce.

For bony fishes and sharks, most of the heat is lost across the skin and gills. Since marine mammals and seabirds lack gills, they suffer little respiratory heat loss. Members of only five bony fish families (including the tunas, marlins, and Swordfish), two shark families (mackerel sharks and threshers), and one ray family (some mantas) have evolved ways of trapping heat and elevating their body (or specific parts of their body) temperatures above that of the environment. The mechanism they employ is elegant and simple and relies on the engineering principle called *countercurrent exchange*. In countercurrent exchange, there will be a near-complete transfer of heat from blood vessels carrying blood warmed near the shark's core to another set of surrounding blood vessels that carry cooler blood from the outer edges of the fish in the opposite direction, to the core. By the time the warm blood reaches the periphery, the heat has been transferred to the cool blood moving toward the core, and thus the core retains the heat.

The evolution of the specialized anatomical structures and physiological processes that enable endothermy are not without explanation; higher internal temperatures allow faster chemical reactions and higher metabolic rates that enable these warm-bodied beasts to be fast-swimming, highly mobile predators—in other words, to maintain a high-performance lifestyle. Endothermy also may allow expansion of an organism's thermal niche, meaning it can move independently of temperature, both latitudinally and vertically within the water column.

Endothermy comes at an apparent cost: the metabolic furnace that powers the high-performance engine of endothermic fishes must be fed. Since the advantages of endothermy almost all relate to creating better predators, until the supply of prey diminishes, fueling the metabolic engine is no problem.

Why Sharks Count

This could be the shortest entry in the book: read the previous entries, which pretty much encompass the ecological, economic, and other reasons for conserving sharks.

Predators help preserve an ecosystem's biodiversity, and they instill fear in their prey, a necessary state that enhances their survival. The economic argument for shark conservation is, namely, preserving ecosystem services (goods and services that are required for human well-being and that ecosystems provide for free) for which there are not sufficient technological replacements. In addition to the biomedical uses and ecotourism we have previously discussed, the kinds of economic services provided by sharks include maintaining biodiversity of marine ecosystems such that populations of commercially and recreationally important fishes continue to exist. Products of economic value that come from sharks include shark meat, fins, leather from shark skin, teeth, and jaws, as well as other products.

That brings us to the intangible reasons why sharks matter. First, sharks provide spiritual and aesthetic refreshment and add an incalculable value to our lives.

Ecologist E. O. Wilson identified an innate human need known as *biophilia* as the urge to affiliate with other forms of life. If you have ever marveled at seeing a shark in nature, or even in an aquarium, then you understand the value that a shark adds to your life. Second, sharks, like every other species on the planet, have intrinsic value; that is, they have a right to exist based on their heritage, and this right is irrevocable.

Given all of the benefits, tangible or not, that sharks and the ecosystems in which they live provide, we ignore their plight at our own risk. Sharks are critical for ocean health, and are simply unequivocally essential for our survival. It thus is incomprehensible that anyone would want to live in a world without sharks, or that we would stand by idly while they suffer the ills of an overcrowded, callous world.

There are some hopeful signs that humanity understands its plight vis-à-vis climate change, biodiversity loss, and so on, and is acting to make corrections that will allow the planet to avoid its worst impacts. We can only hope that it is not a case of too little, too late. Or we can act to make sure that it isn't. You can play a role by leading a sustainable life, educating yourself about the biosphere in general and sharks in particular, educating others, and working to make our current systems of economics and government function for stewardship of the planet, or replacing them, peacefully, with ones that will, for the love of sharks.

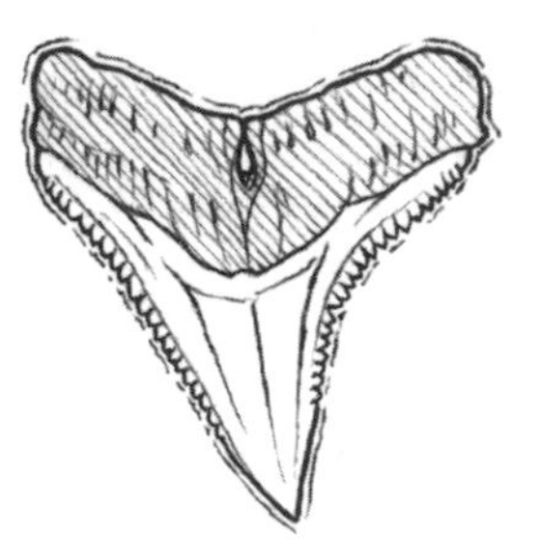

Acknowledgments

Although *Sharkpedia* is diminutive in dimensions, it is broad in the contributions that have culminated in its publication. First, I owe unwavering gratitude to my editor, Robert Kirk, whose tranquil style and inspiring voice enabled me to be a free-range author and plumb the depths of my experience, knowledge, and creativity to write this very cool book about sharks, and to do so with great joy! I also thank the rest of the superlative production team at Princeton University Press, including Mark Bellis, Matt Taylor, Sydney Bartlett, and Megan Mendonça. And a hearty and well-deserved thank you to my superlative copyeditor, Lachlan Brooks, who spared me the embarrassment of alphabetizing mistakes (an unforgiveable offense in this type of book), awkward sentences, and my troublesome, tiresome tendency towards too much alliteration throughout the book.

Sharkpedia represents my first solo effort as a book author, and the coauthors of my previous books played no small role in preparing me for this leap of faith.

They are Robert McConnell, Dean Grubbs, the late Eric Koepfler, Robert Johnson, and Sharon Gilman—scholars, writers, provocateurs, lovers of nature, and dear friends, all.

I'd also like to correct a sin of omission in my previous books in which I failed to acknowledge my undergraduate professors whose skillful instruction and inspiration brought me here: Doctors William D. Anderson, Jim Smiley, Marion Doig, Frank Kinard, Charles Biernbaum, Maggie Pennington, Martha Runey, and Harry Freeman.

When I was offered the opportunity to write *Sharkpedia*, the bait that was used on me was that the book would be illustrated by Marc Dando. If his name is unfamiliar to you, I assure you his magnificent artwork on sharks and other animals is not. I would have, without hesitation, agreed to write a book about the virtues of wicker furniture, of which I know nothing, if Marc Dando was going to be the illustrator. I am in awe of his talent, but even more so of how gracious a person he is.

I would be embarrassed in forgetting to acknowledge others of my remarkable colleagues, students, family, and friends. I am lucky enough for this list to be too long to include their names, and I hope that they do not need this space to know how much I value them. And let me not forget the natural world, including its sharks, that brought me to this point starting when I was in my single digits; it is the sine qua non of our existence.

Finally, I have the privilege of dedicating this book to the senior coauthors of my adult life, my wife Mary and children Juliana and Louis.

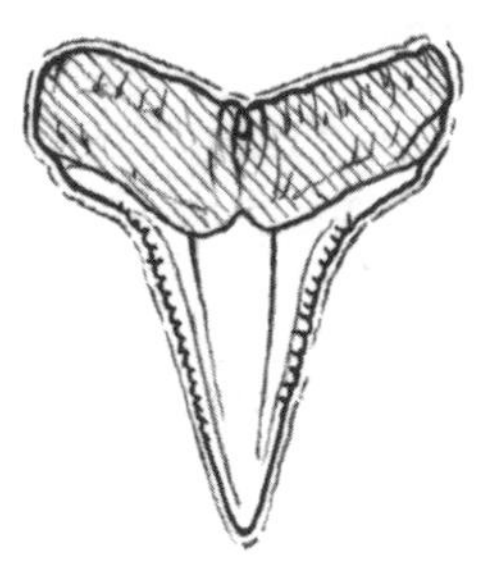

Useful References

In addition to the books listed under the entry "Readings about Sharks," the following studies were consulted in writing this book:

Brown, C., and V. Schluessel. 2023. "Smart Sharks: A Review of Chondrichthyan Cognition." *Animal Cognition* 26: 175–88.

Castro, J. I. 2017. "The Origins and Rise of Shark Biology in the 20th Century." *Marine Fisheries Review* 78: 17–28.

Gallagher, A. J., and N. Hammerschlag. 2011. "Global Shark Currency: The Distribution, Frequency, and Economic Value of Shark Ecotourism." *Current Issues in Tourism* 14: 797–812.

Gulak, S.J.B. et al. 2015. "Hooking Mortality of Scalloped Hammerhead Sphyrna Lewini and Great Hammerhead Sphyrna Mokarran Sharks Caught on Bottom Longlines." *African Journal of Marine Science* 37: 267–73.

Haas, A. R. et al. 2017. "The Contemporary Economic Value of Elasmobranchs in The Bahamas: Reaping the Rewards of 25 Years of Stewardship and Conservation." *Biological Conservation* 207: 55–63.

Hayes, E. et al. 2018. "Basking Shark Breaching Behaviour Observations West of Shetland." *Marine Biodiversity Records* 1: 1–5.

Klimley, A. P. 2023. "A Historical Approach to Describing the Complex Behaviour of a Large Species of Carnivorous Shark. Case Study No. 1: The Scalloped Hammerhead, Sphyrna Lewini." *Behaviour* 1: 1–22.

Moss, S. A. 1972. "The Feeding Mechanism of Sharks of the Family Carcharhinidae." *Journal of Zoology* 167: 423–36.

Pratt, H. L., and J. C. Carrier. 2001. "A Review of Elasmobranch Reproductive Behavior with a Case Study on the Nurse Shark, Ginglymostoma Cirratum." *Environmental Biology of Fishes* 60: 157–88.

Wood, A. D. et al. 2009. "Recalculated Diet and Daily Ration of the Shortfin Mako (Isurus Oxyrinchus), with a Focus on Quantifying Predation on Bluefish (Pomatomus Saltatrix) in the Northwest Atlantic Ocean." *Fishery Bulletin* 107: 76–88.

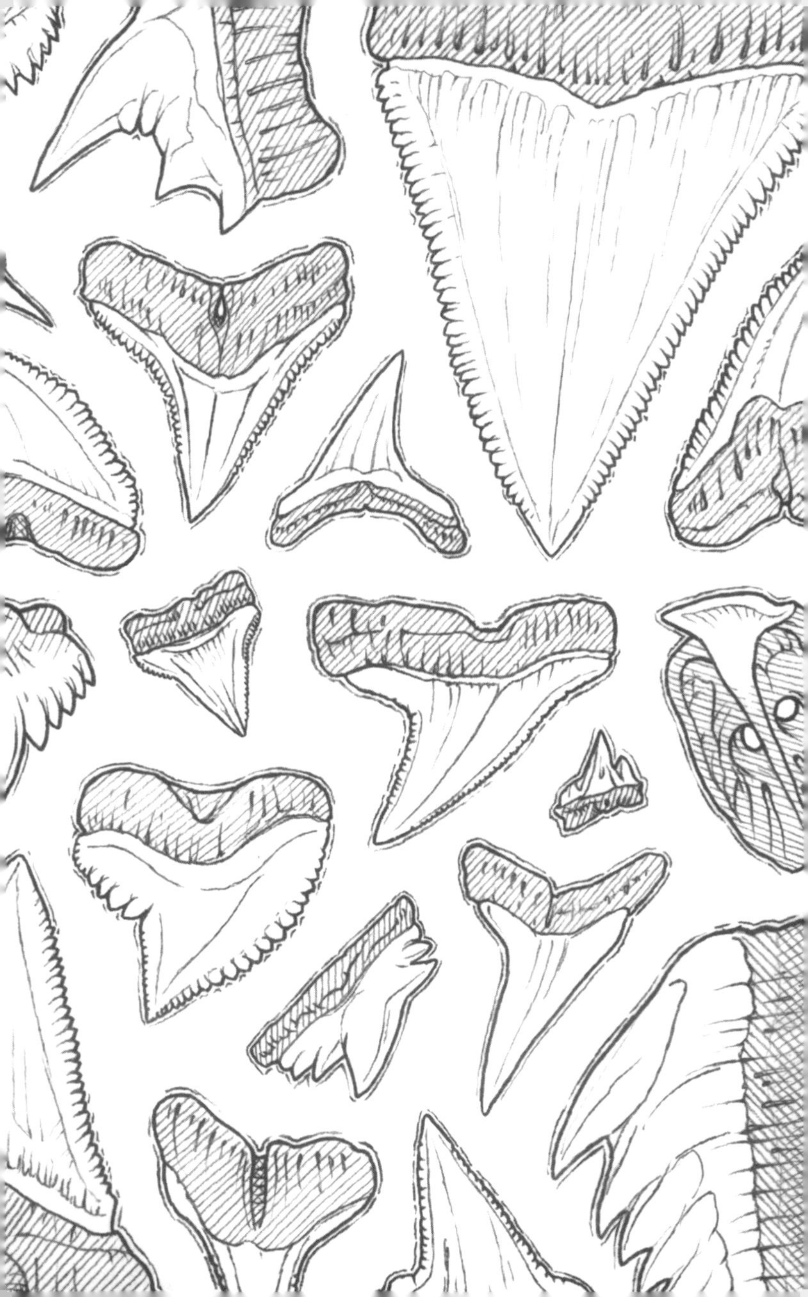

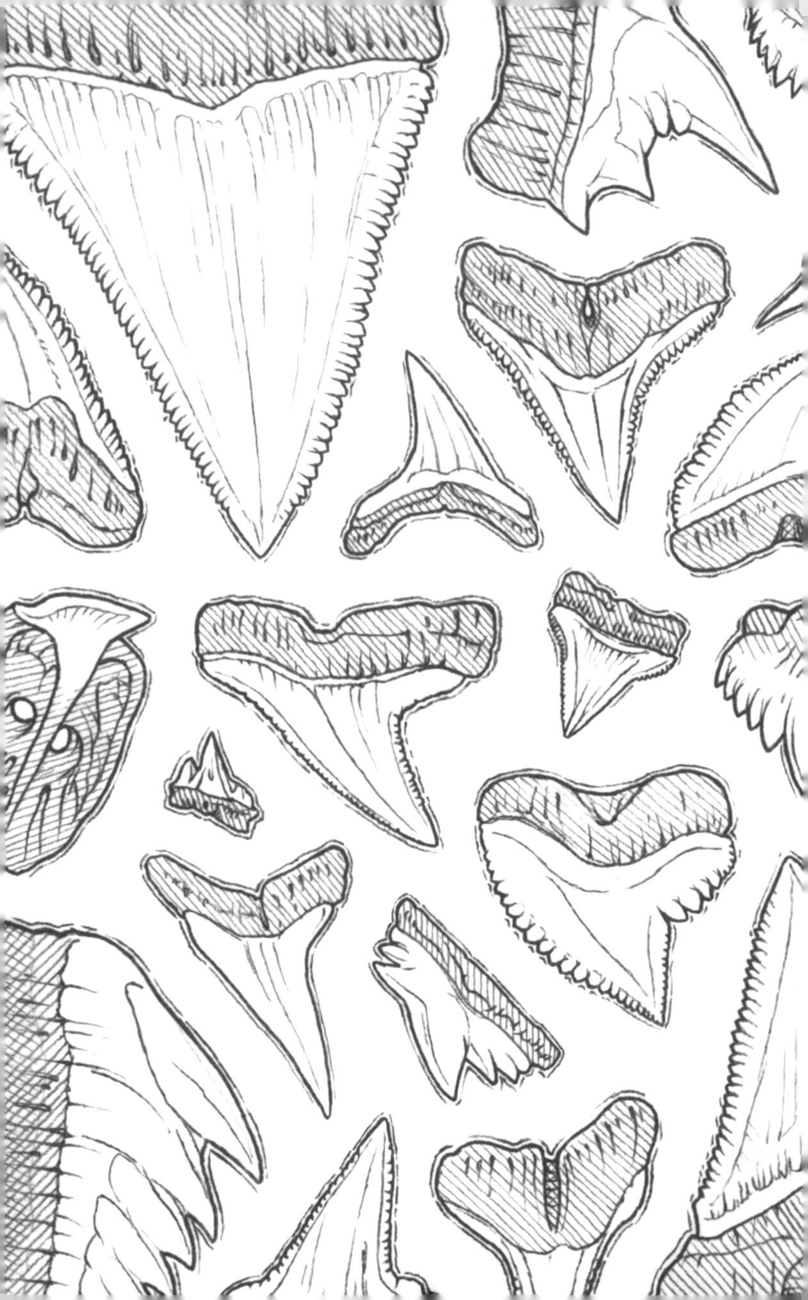